W0037979

UNNATURAL
TEXAS?

GIDEON LINCECUM NATURE & ENVIRONMENT SERIES

Sponsored by Jerry B. Lincecum and Peggy A. Redshaw

UNNATURAL
TEXAS?

THE INVASIVE SPECIES DILEMMA

Robin W. Doughty & Matt Warnock Turner

TEXAS A&M UNIVERSITY PRESS • College Station

This paper meets the requirements
of ANSI/NISO Z39.48–1992 (Permanence of Paper).
Binding materials have been chosen for durability.
Manufactured in China by Everbest Printing Co.
through Four Color Print Group

LIBRARY OF CONGRESS CATALOGING-IN-PUBLICATION DATA

Names: Doughty, Robin W., author. | Turner, Matt Warnock, 1960– author.
Title: Unnatural Texas? : the invasive species dilemma / Robin W. Doughty and
    Matt Warnock Turner.
Other titles: Gideon Lincecum nature and environment series.
Description: First edition. | College Station : Texas A&M University Press,
    [2019] | Series: Gideon Lincecum nature and environment series | Includes
    bibliographical references and index.
Identifiers: LCCN 2018028140| ISBN 9781623497057 (book/cloth) | ISBN
    1623497051 (book/cloth) | ISBN 9781623497064 (e-book)
Subjects: LCSH: Introduced organisms—Control—Texas. | Biological
    invasions—Environmental aspects—Texas. | Ecological disturbances—Texas.
    | Ecology—Environmental aspects—Texas. | Biodiversity
    conservation—Texas. | Nature conservation—Texas.
Classification: LCC QH353 .D68 2019 | DDC 578.6/2—dc23 LC record available at
    https://lccn.loc.gov/2018028140

CONTENTS

PREFACE

The Charles Alan Wright Fields (formerly known as Whitaker Fields) is an intramural facility in Austin owned by the University of Texas. Today, semester finals are finished, and except for a single jogger cantering along the perimeter, the 34-acre sports field is empty. A mower buzzes inside the fence, readying soccer and football pitches for future fixtures.

However, noon in mid-May is more for birds than athletes. There are parrots fluttering and squawking among the light stands 30 or so feet above the ground. The sturdy metal fixtures bear bulky nests. The parrots have chosen well. There is no better site—too high for opossums and raccoons to climb and so enmeshed in electrical bric-a-brac that a hawk would find it hard to pick off an exposed parrot. Chambers in the woody thatch conceal birds.

The green-backed, gray-breasted parrots just under a foot long have woven stick after stick to make half a dozen or more woodpiles. Someone has claimed discovering a nest as large as a VW Beetle. Called monk parakeets, the

Monk parakeet carrying bulky nesting material. Photo by Manjith Kainickara (CC BY-SA 2.0).

Monk parakeets building nests on transmission poles at Charles Alan Wright Fields in Austin, Texas. Photo by Matt Turner.

midsize species is native to Argentina and nearby countries in South America. As the name suggests, the somewhat drab-looking birds do not sport cherry-red heads or graceful silhouettes of the red-masked parakeets on San Francisco's Telegraph Hill, made famous by the similarly titled film. This is a less showy species, also called quaker parrots, reportedly because of young birds' habits of trembling or quaking while being fed. They are flourishing in Austin, and, like their Bay Area counterparts, capital-city Texans enjoy them. More than 50 monk parakeet colonies exist in Austin and are sprouting in upscale neighborhoods. You can sip a latte in Tarrytown's Casis Shopping Center and enjoy the squawks and antics of 30 or so parakeets that have commandeered a platform on a nearby shortwave tower. People released some parakeets in the 1960s, and today populations of this adaptable urban species reside in 15 or more US states and in several European nations. This gregarious bird is part of what makes Austin "Weird."

Flying about Wright Fields, the parakeets drink from leaky faucets and feed acrobatically in nearby trees on berries and buds. Monk parakeets are garrulous and give you a mischievous stare as befits their intelligence. There are other birds too. A northern mockingbird, Texas' state bird, swoops on the turf to snap up a bug. Barn swallows scythe back and forth above black-garbed great-tailed grackles that waddle on the grass. These grackles have multiplied in Austin and throughout Texas during the past 80 or 90 years, adapting to suburbs with lawns and mown spaces, such as Wright Fields. They are native, although the northern edge of their range barely reached south Texas in 1900. Seeds and insects on suburban lawns, irrigated campuses, parks, and croplands have lured them northward to be one of North America's successful bird species. Around the University of Texas campus, grackles have also adopted a habit of dumpster diving and pecking scraps from encrusted food containers behind fast-food outlets. Not all people have kind words for these wheezy-voiced strutters. Relying on firecrackers and blanks from shotguns, university personnel banished them off campus through downtown and into the city's Zilker Park, where thousands gather to roost in winter. Nearby residents hate them for fouling vehicles and, when they have young in a nearby nest, occasionally pecking at people. Other Austinites embrace their big, Hitchcock-esque gatherings, and grackle images appear on paintings, T-shirts, and socks. A 37-foot kinetic sculpture resembling a cuckoo clock with blackbirds in Austin's new flagship public library was inspired in part by the city's grackles.[1]

There is another bird that is just as dark in plumage and even darker in reputation. At least 50 nonnative European starlings are probing newly mown play areas. The European starling is one of the most despised birds in North America. It damages fruit orchards and vineyards and competes for nest houses put up for native bluebirds, martins, and titmice.

Enthusiasts introduced Old England's starlings into New England more than a century ago, and within 50 years the scoundrels have flown from New York City, their release point, to Beaumont, Texas. Most people do not recognize starlings, as they do parakeets, and do not know what pests they have become. But farmers and bird experts do, knowing how early promoters failed to take into account the bird's natural history before releasing it on a new continent.

The starlings scatter off the shorn Bermuda grass, an introduced hybrid grass, and fly off to surrounding trees and bushes. Like the parakeets, their presence shows that Austin, like other American cities, owns a mix of native and nonnative animals and plants. Austin's parakeets and starlings have additional alien cousins, such as the house or English sparrow, as well as scores of nonnative ornamental flowers, shrubs, bushes, and trees, including Chinese tallow, that flourish in various neighborhoods. Every year plant nurseries in Austin, throughout the state, and across the nation market nonnative flowers, pond plants, shrubs, and trees to fashion-conscious homeowners, who are looking for an accent, maybe a splash of color or something new and trendy for their yards. Most of these foreign species die off, but some hang on, and several start to flourish and spread beyond the beds and gardens where green-thumbed owners planted them.

We are only just beginning to understand the interactions of this mix and match of organisms inside modern cities; some nonnative plants jump fences and spread along riverbanks, alleyways, and industrial sites; some choke out native vegetation and are condemned for it; others have remained more or less localized or then have grown noticeably populous and compete with native species. Japanese knotweed, from Asia, as its name suggests, has heart-shaped leaves, pretty white flowers, and supple stems but has earned a reputation as obnoxious in the United States. It is reputed to bust through pavements and sidewalks, even into buildings' foundations. So far, the knotweed has conquered 29 nations, including 42 US states and eight Canadian provinces during its 120 or so years in North America. It is one of two foreign plants declared illegal to grow in the United Kingdom, where it was introduced in about 1840. Sold as an ornamental and also for erosion control, knotweed takes over gardens and yards. Dense 15-foot-tall thickets useless for wildlife spread beside watercourses. According to invasive plant specialists, this tough plant is standing on the doorstep of Texas.[2]

Knotweed is a bit like the native grackle in Austin, which is not officially called invasive because it is a native bird. Both knotweed and the grackle have adjusted well to urban and suburban locations. Ninety years ago grackles were virtually unknown in Central Texas; today they are among the better-known Austin birds. They have moved inland as far north as southern Nebraska and as far west as south-central California.

It is not always clear why one native or nonnative plant or animal thrives while another dies out. Evidence suggests that both the alien knotweed and native grackle prefer disturbed and transformed sites, making opportunistic use of cleared surfaces, exposed soils, new habitats, a range of climatic conditions, and available resources that exemplify the dynamic nature of urban ecology. While some experts champion the cause of native animals and plants and push for control of nonnative species, especially ones that have become too successful (usually meaning harmful), other experts argue that nonnative and native organisms are coexisting and that some introduced species may serve a purpose in the ecosystem—buffering against storm events; furnishing cover, breeding spaces, sources of nectar, buds, seeds, and berries for native insects, birds, and mammals; or, like the parakeets, just being a pleasure to have around. Austin Energy crews have taken an active stance by pulling down parakeet nests (some with eggs or young) along city streets on the grounds that the sticks pose a fire hazard or may interrupt power to users. A video caption of a parakeet nest catching fire declared, "Power knocked out to 1,000 customers in Austin." Bird lovers were appalled by this treatment and urged the energy authority to wait until after the breeding season. A truce was declared. This incident shows how quite suddenly a seemingly friendly foreign species can attract sufficient bad press to turn it into a menace to a city's economy or public health. The parakeet is still the same bird, but some of us have started to characterize it as a pest and treat it as such.[3]

This book explores what has motivated both these sorts of attitudes and treatment of a selected number of introduced plants and animals in Texas. We focus on several well-known species that most people now regard as invasive. By definition, an invasive species is a nonnative plant or animal that has become too harmful for our liking, so we have devised programs and procedures for controlling, or even eradicating, it. This attitude toward unwelcome species mirrors the very successful antilittering campaign, "Don't Mess with Texas." By introducing nonnative plants and animals, we have "messed" with Texas, although for a long time we did not believe that was the case.

Early settlers took it for granted that the plants and animals they took with them and released "improved" the local flora and fauna. They depended on domesticated plants, livestock, and pet animals for food, recreation, decoration, and companionship. Planting corn and cotton, running cattle, goats, and pigs were usual, tried and tested aspects of settlement. However, Texans also developed an appreciation for native species, listing the bluebonnet, mockingbird, and pecan tree as the state's flower, bird, and tree, respectively. Having moved into Texas from Mexico in the 1800s, the nine-banded armadillo is now the state's official small mammal, and the Mexican or Brazilian free-tailed bat its official flying mammal.

Texans are proud of their biological heritage and enjoy seeing plant and animal spectacles, such as the color of fall foliage in Lost Maples State Park or the flight of bats under Austin's Congress Avenue Bridge. There are now more bird festivals, native plant conferences, and workshops devoted to growing drought-tolerant plants and xeriscaping than ever before. This preference for native species means less tolerance for nonnative ones, especially for those that we believe do damage to the economic health, well-being, and integrity of local landscapes.

Texans have learned the hard way that some of these foreign species have not only damaged the state's economy but also affected its ecology. Today, we want to understand costs as well as benefits associated with introducing plants and animals from different regions of the world, because we have discovered that some species that our forebears regarded as beneficial are now labeled as pests. This book shows how our affection for these species has changed and discusses taking responsibility for what has happened, not to blame earlier generations for maximizing resources in Texas by introducing new species but to make sure we do not repeat earlier mistakes. Earlier Texans had no doubts that bringing in potentially useful or interesting plants and animals was good for making a home here. What we have learned is that what was regarded as a resource can come to be a pest. Therefore, it is important to challenge assumptions about what a species does, or will do, and not ignore or discount indications or evidence that an organism we have introduced is causing problems or is likely to do so.

ACKNOWLEDGMENTS

We especially want to thank the following individuals whose time, energy, and insights helped shape this book: Damon Waitt, founder of the Texas Invasive Plant and Pest Council and member of the federally chartered Invasive Species Advisory Committee; Larry Gilbert, director of the University of Texas at Austin's Brackenridge Field Laboratory; Rob Plowes, research scientist and codirector of the lab's invasive species program; and Kevin Anderson, a geographer who runs the Austin Water Utility, Center for Environmental Research.

We are grateful to Texas Parks and Wildlife officials for the latest updates and perspectives: to John Davis, director of the Wildlife Diversity Program; Earl Chilton II, director of the Aquatic Invasive Species Program; Kelly Conrad Simon, urban wildlife biologist; Donnie Frels, project leader for the Edwards Plateau Ecosystem Management Project; Richard Heilbrun, leader of the Conservation Outreach Program; and Dale Prochaska, regional director of the Wildlife Division.

Several officials with the Texas Department of Agriculture deserve our gratitude for offering insights into how invasive organisms affect our agricultural and health: Awinash Bhatkar, Janet Fults, Allison Olofson, and David Villarreal.

Colleagues in the Department of Geography and the Environment at the University of Texas in Austin, especially Ian Manners and Paul Adams, have been most helpful in fostering research about this topic.

There are many people who strengthened and enriched this work. Our thanks go to John Abbot, Paul C. Adams, Heather Allard, the late Tony Amos, David Bamberger, Bob Barth, Amy J. Benson, Paul Bonin-Rodriguez, Charlotte Boyle, Vaughn M. Bryant, Rob Denkhaus, Saara DeWalt, Tim Eaton, Bill Edwards, Nancy Elder, Beverly Garland, Cullen Hanks, Shane Harrington, Simon Humphrey, Darrell Hutchinson, Harold Jobes, Laura Joseph, John Kinsey, Robert Knight, Norman Kutcher, Edward LeBrun, Ian Manners, Marsha May, Jerry McDonald, Jon Mooallem, Jonathan Ogren, Lars Pomara, Barbara Porter, Monique Reed, Tiana Rehman, Chuck Sexton,

Beryl Simpson, Judith Sims, Judith Stamper, Tom Stehn, David Todd, Billie L. Turner, Nina Leigh Vizcarrando, Richard Wallach, and Chip Wozniak.

Finally, we deeply appreciate the people at Texas A&M University Press, Patricia Clabaugh, Shannon Davies, Stacy Eisenstack, and Cynthia Lindlof, whose foresight, faith, and patience have brought this book to fruition.

UNNATURAL
TEXAS?

INTRODUCTION

You can eat tomatoes in Italy, hunt oryx in Texas, ride horses in Chile, curse cane toads in Australia, dig earthworms in eastern North America and catch rats in the Galápagos—all are part of the global homogenization of flora and fauna.

—Jeremy Hance, "How Humans are Driving the Sixth Mass Extinction"

Why should we in North America bother about introduced starlings as they waddle after grubs in our lawns and peer into birdhouses as possible nest sites? Should we be concerned that stands of nonnative Chinese tallow trees are springing up on the coastal grasslands of Texas or that another woody import, saltcedar, takes root in draws and river bottoms farther west? Does it matter that Asian grass carp, introduced to control the pesky water weed called hydrilla (also from Asia), are also chomping on native aquatic plants in our lakes or reservoirs? Should we mind that free-ranging herds of elegantly spotted axis deer from India, Nepal, and Sri Lanka are bounding across openings and slopes with our native white-tailed deer? To what extent does all this global intermingling of plants and animals concern us? We have been altering the natural patterns of flora and fauna across the globe for a long time and are just speeding up the process.

There are several responses to having increasing numbers of nonnative species around us. One of them seemingly adopts the British meme "Keep Calm and Carry On" used on posters during World War II. The catchphrase means relax and do not worry, because like any animal, the starlings, Asian carp, and Asian deer have the right to live the way they do. Supporters of People for the Ethical Treatment of Animals (PETA) and similar groups aimed at treating animals humanely argue that, native or not, animals have the intrinsic right to follow the ways their evolutionary paths have led them. The fact that some misguided persons released sparrows, nutria, or lots of different fish, including the European carp in Texas long before a similar species arrived from Asia, is not the fault of the animals. They were not responsible for this ongoing and accelerating worldwide movement and buildup of plants and animals; we are. And the fact that some exotic game mammals are endangered in their homelands gives moral weight to having them in Texas. Finally, many creatures, like feral hogs and feral cats, have been here for hundreds of years, and many regard them, even in a wild state, as part of Texas' fauna, even if they were outsiders at one time. How far back do we have to go to categorize organisms as native or nonnative? Perhaps, therefore, we should allow these animals and others to exist where they have become established. Being alive is its own reward.

At the other end of the nonnative acceptance spectrum are people who want to preserve and retain natural ecosystems as they existed before all this exchange happened and, if possible, restore them. They tend to draw a firm distinction between what is native and what is not. They recognize that animals, plants, and microorganisms have come together in given locales and have formed intricate relationships with one another and with the environment that, broadly speaking, bring about a kind of balance. The sudden intrusion and multiplication of organisms alien to these ecosystems offsets that balance, and the diversity that characterizes and supports it, sometimes radically. These biological nativists argue that flora and fauna that have evolved in North America, for example, or moved in without human intervention deserve protection and management when faced by threats. Conversely, nonnative species—ones that we have introduced by choice or inadvertence, such as the European starling and zebra mussel, should be respectively controlled and eradicated if possible. Many state wildlife management policies and programs adopt this position, preferring native species to introduced ones, even though an animal or plant may be long established and in a sense naturalized in a state or region.

Between the two ends of this spectrum are positions that are more pragmatic. Many experts, for example, argue that although it is impossible to eradicate entrenched nonnatives, such as the house sparrow in Texas and North America, it is possible to control problem-causing groups or populations. We can get rid of sparrows, which damage cavity-nesting native wrens, bluebirds, and similar avian favorites, and starlings, which strip vineyards, peck orchard fruits, and transmit diseases. This pragmatic stance supports trapping the alien birds in certain localities or at certain times: around birdhouses during the nesting season, for example, where both sparrows and starlings compete with and prey on purple martins. The purple martin, America's largest and much-loved swallow, relies almost exclusively on nest cavities supplied by human well-wishers. Many so-called martin landlords evict or even kill sparrows and starlings that invade their colonies.

Invasive species, the subject of this book, are nonnative organisms that cause explicit harm to the economy, human health, or environment. They actually make up only a fraction of all nonnative species, but because of their abundance and the disproportionate impact they may inflict on the environment, and us, they rightfully grab the lion's share of our attention. This book uses Texas as its focus. Although every invasive species mentioned inhabits other states and may occur in other nations as well, the combination and relative abundance of the species are unique to this state. We want to take on some of the Lone Star State's biggest hitters, so to speak, that is, those that are widespread and cause the most damage.

This book explores two things. First we discuss some of the better-known invasive organisms in Texas. They are among the most widespread and injurious to the state's economy, its land and waters, and the health of its citizens. We ask how different invasive species arrived in Texas and what these "New Texans" have done and are doing to the Lone Star State. The answers to these questions are largely about facts. They are the purview of science: biology, ecology, geography, agricultural science, and similar disciplines. They are what most people expect of an invasive species book.

But we go beyond this guidebook approach by exploring the subjective world of attitude, culture, and belief that enriches and complicates the documented facts about invasive species. This realm of human attitude and behavior helps explain why we introduced plants and animals in the first place, then neglected or forgot about some, while shifting opinion from approval to disapproval about others. Newcomer species remind us how often we take our relationship to nature for granted. Of course, we all agree about the need to use nature; we could not feed, clothe, or shelter ourselves otherwise. But how to use it and in what ways causes debate, especially when we assume that we have the right, duty, and competence to "improve" nature, making Texas "better" than it once was. Only recently have we considered and come to accept there is value in leaving Texas' nature alone, enjoying its natural vitality and character. We recognize there are limits to what we can know or do know.

There is often a push and pull between facts and what we make of them. Life scientists and ecologists may agree on what constitutes an invasive organism, but the general public may resist such hard-and-fast distinctions. Widely held attitudes toward plants and animals in this state and across the nation change, so an organism may be regarded as an invaluable resource in one place and then be rejected as a pest in another time. Harm is at the very core of the term "invasive," yet to what extent harm happens is subjective, especially in light of the fact that nonnative species may also provide beauty, resources, and ecosystem functions. Ethical and highly emotional overtones arise with certain species that also affect how we go about controlling their numbers and spread. Thus, our understanding of an invader species, how it adapts to new conditions, and the appropriate methods for dealing with it remain inexact, often fragmentary, and open to revision.

Nonnative plants and animals reveal our limitations and shortsighted-ness and also demonstrate a paradox that this book exposes: there is value in inaction as well as action, in nonuse as well as use, in nonconsumption as well as consumption, of having things around to observe and enjoy, like the monk parakeet in Austin, Texas, so watching and waiting are neces-sary before seizing on all means to eradicate it or charging in before under-standing the niche it is using or role it is playing. In the 1970s more than 1,000 pounds of live Louisiana crawfish were released for aquaculture in the Guadalquivir River marshlands of southwestern Spain, where they became established and are now considered one of the worst nonnative species in Europe. However, in 2013, a move to list the alien crawfish as an invasive species threatened local jobs tied to the sizable export industry that had developed as a result of its abundance, demonstrating how the political and economic overtones may shadow the biology of an unwanted species unless collaboration is achieved.[1]

Are there cases where humanly induced changes are so indelible that there is neither a going back nor a putting back? Authors of a study of an urbanized wetland in Taiwan discovered that nonnative fish species, notably tilapia, are the most important food items for migratory and resident water-birds. They concluded that as substitutes for native species, the introduced fish "tolerate severe human-disturbances and exert beneficial effects on the ES [Environmental Services], so they could be temporarily retained to facil-itate ecological restoration of the urbanized wetland ecosystems," or in similar ones, such as city wastewater treatment plants that are constructed from start to finish. Closer to home, wildlife officials found that the endan-gered California clapper rail was just as happy to nest in an invasive hybrid cordgrass, targeted for eradication, as it was in the native cordgrass that the hybrid had largely replaced around San Francisco Bay. Removal of the invasive grass was actually causing a decline in the bird's numbers, so offi-cials altered their strategies to accommodate habitat restoration. In the case of both fish and cordgrass, a simple all-out removal of the invasive species would arguably cause more harm than good.[2]

Stories about invasive species open our eyes to past priorities and poli-cies, to fads and assumptions, to mistakes both innocent and culpable—all of them will forever dog us. The point of the book is to make us more aware of the effects we have had and continue to have on the face of Texas, how zealotry and single-mindedness may ignore or gloss over problems, and how important it is to take more than a species approach in policy and decision making by including a broader and more holistic view that is proactive rather than crisis driven. It is a judgment about native and nonnative species that is less professionally compartmentalized, politically judgmental, and socially divisive. As Banu Subramaniam notes, instead of placing scarce resources into checking political boundaries and borders, it is better to take seriously unchecked development, inadequate environmental laws and regulations, and the globalization of plants and animals for purely economic ends. "Ultimately, the campaign against the foreign does not solve species extinctions or habitat degradation," she insists.[3]

This does not mean that invasive species do not cause problems. They do, and tremendous ones at that, but they are not exclusively to blame for the ongoing loss of diversity and threats posed to native species. Before coming to grips with problem plants and animals in Texas, we first need the general context on how species have traveled around the world to get here and the unforeseen consequences of their exchange. This includes vocabularies we have coined about what an invasive species is and how it tends to behave.

Worldwide Species Exchange

The term "Anthropocene," or Age of Humans, is currently being debated concerning whether it should be formally recognized as a new geological period. The term refers to human population growth, spread, and activities that have resulted in massive and continuous transformations of lands and waters across the surface of the planet. Some declare that the invention and expansion of agriculture sparked the Age of Humans. Another suggestion refers to its beginning in the early 1600s when the Old World and New World were joined irrevocably by exploration and conquest. This included shipments of plants and animals (mostly domesticated ones) as well as people across the Atlantic Ocean. Or, even more recently, this beginning of intense activity and lasting impression is perhaps best defined by nuclear tests that led to worldwide radioactive fallout in the 1950s and 1960s. Geologist Colin Waters, secretary of the working group investigating the Anthropocene boundary, insists, "Being able to pinpoint the interval of time is saying something about how we have had an incredible impact on the environment of our planet."[4] Thus, all starting points for this Age of Humans involve ways we have exploited, manipulated, and dispersed resources and spread all sorts of species among continents and oceans.[5]

Geographer Alfred Crosby characterizes this switch of flora and fauna between Western Europe and the Americas as the Columbian Exchange. Food was a key reason for our habit of redistributing plants and animals, he notes. For example, on arrival in 1607, colonists in Jamestown cleared land to plant wheat and fashioned a garden in which to set fruit and vegetables "not indigenous to their new home." Some years later, the Virginia Company dispatched a pinnace to Jamestown loaded with wheat, barley, garden seeds, and the scions of fruit trees. It was clearly the thing to do so that colonists would feel at home in a new land—make it resemble Europe.[6]

Almost all of our current diet depends on what the Columbian Exchange has given us. It has meant transporting, planting, cultivating, and marketing mostly domesticated plants and animals to satisfy our daily needs. Columbus loaded cattle, pigs, and sugarcane for his second crossing in 1493 and sailed back to Spain with corn (maize), peppers, and pineapples. Twenty years after that, Hernán Cortés carried the turkey, reportedly domesticated by the Aztecs, from Mexico to Spain, only to have it passed to other nations, including England by 1541. English colonists reintroduced carefully bred toms and hens along the Eastern Seaboard: some were put ashore in Jamestown, Virginia, in 1608. European nations exploited and transported biota, as well as indigenous people, they considered useful to far-flung and remote areas. They often assumed that native plants and animals in those areas were inferior and were less beneficial than the familiar ones they carried in and distributed. It was also a matter of ignorance: not caring about using indigenous species when a suite of tried-and-tested European houseplants and farm animals sufficed.

Often with government backing or tacit approval, settlers, landowners, and entrepreneurs imported a veritable Noah's Ark of crop and ornamental plants, including shrubs and trees, domesticated livestock, poultry, plus game mammals and birds, songbirds, house pets, and, of course rodents and insects (which made crossings in goods and foodstuffs and inhabited most ships).[7] In addition, so-called acclimatization societies arose to encourage the introduction of nonnative species to improve the supplies and range of foods. These organizations established in the 1800s experimented with shipping animals from temperate regions in the Northern Hemisphere to similar regions in the Southern or Western Hemisphere. Potential domesticates, such as the common eland, were placed on steamers sailing out of southern Africa for Europe. In some cases, these transfers reflected strong cultural prejudices. For instance, enthusiasts shipped British and European plants and animals to Australia and New Zealand to enliven, they claimed, what was a dull and empty landscape and odd-looking fauna. New Zealand, for instance, barely had any land mammals at all, so it was going to be made another England far away in the south.[8]

In the other direction, eucalyptus trees grew to be a popular export item from Australia. Acclimatization groups believed the hardwoods would make excellent building materials, especially after they grew rapidly around the Mediterranean Sea (and in California), and would purify the air of disease-causing miasma. Colonial and world exhibitions celebrated and promoted the trees. So-called gum tree boosters planted seeds widely in the Northern Hemisphere, including French-held North Africa, British-held India, and other regions with a Mediterranean climate, such as in Spain, Italy, and California. They also planted them successfully as far north as Scotland.[9]

In the United States, at least half a dozen acclimatization societies appeared after 1850. Located in New York, the most famous of them was led by amateur bird enthusiast and pharmacist Eugene Schieffelin, who among other things, through his organization called the Friends of Shakespeare, dedicated himself to releasing all the birds named in the works of the Bard of Avon. Schieffelin presided over the American Acclimatization Society, founded with fellow enthusiasts in 1871, to introduce useful or interesting foreign animals and plants into the United States. His work resulted in the European, also called common, starling being released in New York together with Japanese finches in 1890 and repeated in 1891. By that time, the English or house sparrow had become well known in New York State, having gained a foothold in 1853 after individual sparrows shipped in from Liverpool, England, had begun to nest in the 478-acre Green-Wood Cemetery, Brooklyn, founded 15 years earlier and increasingly popular among sightseers.

Schieffelin's fad for "improving" American birdlife proved contagious. Cambridge, Massachusetts; Portland, Oregon; San Francisco, California; Cincinnati, Ohio; and even the nascent state of Hawaii followed the acclimation bent for importing nonnative birds. It was a way of selectively increasing useful, beautiful, and agreeable species, especially songsters, proponents noted. It was also a way of reminding Anglo immigrants about their European roots. Both older and younger foreigners, it was claimed, habituated to the chirps of house sparrows, would welcome the same sounds from the same birds along the US Atlantic Seaboard. Amid the bustle and clatter of the new nation, sparrows and other species were expected to be reminders of the Old Country.[10]

Governments were also formally involved in the great swap of organisms around the world. The US Department of Agriculture (USDA) established a Section of Seed and Plant Introduction in 1898. Under the direction of

David Fairchild and placed under the Bureau of Plant Industry, it acquired renown over the ensuing decades for exploring the far corners of the world for new (especially cold- and drought-hardy) cereals, forage, fruit, and ornamental plants. We have these efforts to thank for such additions to US agriculture as the navel orange, durum wheat, Egyptian cotton, the date palm, and new varieties of soybean. Ornamental shrubs and trees, to judge from one of the bureau's annual published lists of new and rare plants, include such exotics as honeysuckles from Tibet, rhododendrons from Manchuria, several species of eucalyptus and privet, and the Brazilian peppertree.[11]

Unforeseen Consequences

A darker side eventually emerged from all this exuberant interchange, which is essentially the subject of this book. Too much of a good thing can become indigestible. Several species of European and Chinese privet (*Ligustrum* spp.) are some of the most widely spread invasive plants in the US South, forming dense thickets in bottomland forests where they replace native shrubs and

Wintering cedar waxwings feast on fruits of the exotic glossy privet *(Ligustrum lucidum)* in Austin, Texas. The tree-sized evergreen, introduced to the United States from Asia around 1800, was widely planted as an ornamental in the South. Its appeal in the nursery trade and general absence from cropping areas tend to keep it off state invasive lists. Photo by Paul C. Adams.

mid-canopy trees. The volatile oil in eucalyptus turns out to be highly flammable, and planted woodlands in California have turned into firestorms, such as the Bay Area's Oakland Hills fire that destroyed almost 3,000 homes and killed 25 people in 1991. The Brazilian peppertree, a recently introduced ornamental plant, is now considered invasive in coastal Texas and is prohibited from import, sale, or distribution in the state. Chinese tallow—an earlier promotion of the bureau and also now considered invasive and prohibited from sale—is so controversial in Texas that we give it its own chapter.

Animals exhibit the same problems. The Burmese python, introduced from Southeast Asia to gain newfound freedom in Florida's Everglades in the 1980s, has completely disrupted the structure of the wetland community. After its introduction, sightings of raccoon, opossum, bobcat, rabbit, fox, and other mammals have plummeted by 90 percent or more, and the snake has also fed on herons and storks.[12]

House sparrows have become ubiquitous throughout the nation and molest 70 species of native birds. One of Schieffelin's darlings as homage to the bard, the European starling, has exploded into enormous flocks that damage crops, spread *Salmonella* among feedlots, and drop vast amounts of poop on city buildings, pavements, and trees.

Insects, though tiny, manage to create real crises as well. The grape phylloxera, a sap-sucking louse native to America, found its way to France in the 1860s, where it wreaked havoc on French vineyards. Native American grapes were largely resistant to this insect, but their European cousins were not. Phylloxera flourished abroad, girdling rootstocks throughout Europe. Over the course of two decades, two-thirds of the continent's vineyards were destroyed, including practically all of those in southern France. Many vintners lost their livelihood, and at least one-third of France's vineyards were never replaced. The only thing that saved the wine industry was grafting European vines with rootstocks that were hybrids with American (especially native Texan) vines. Denison resident and grape expert Thomas Volney Munson deserves credit for saving the French industry in the 1880s for which the Legion of Honor Chevalier du Mérite Agricole was conferred on him.[13]

A century later from the Old World to the New World came the Mediterranean fruit fly. In 1975, the first so-called medflies, originally from Europe's Mediterranean basin, were identified in California. Agencies sprayed insecticide to control this pest that lays its eggs under the skin of fruit. Despite these measures, the medfly population exploded again in 1989 and threatened to ruin the state's citrus industry. In the 20 years after its first appearance, officials shelled out at least $170 million to deal with the medfly in

California, one of three states (Texas is one) that have undertaken campaigns to control and eliminate this rapid colonizer and global menace. The Mediterranean fruit fly now inhabits most of Africa, the Middle East, and large portions of Central and South America. It has even reached parts of Western Australia. The medfly still lives in the Golden State (and Florida), and from time to time outbreaks occur, perhaps because of accidental reintroductions or boomlets in entrenched populations. The last occurrence in Texas was in 1966; however, there is a high risk of outbreaks.[14]

These kinds of problems posed by nonnative organisms are becoming increasingly common. Once an insect, like the medfly, gets established, it is often difficult to control and impossible to eliminate because many nonnative species have no natural predators in the place where they turn up. So they are able to multiply and spread. Some make their presence felt almost immediately and infest crops, livestock, and native species. Others lie dormant, so to speak, for years before suddenly ballooning into serious pests. By no means do all nonnatives behave the same way.

Mediterranean fruit fly. Photo by Florida Division of Plant Industry, Florida Department of Agriculture and Consumer Services, Bugwood.org.

Definitions and Characteristics

Before going further, we need to be distinct about a few critical terms and their meanings.[15]

- **Ecosystem:** a community of organisms and their physical environment
- **Introduction:** the intentional or unintentional escape, release, dissemination, or placement of a species into an ecosystem as a result of human activity
- **Invasive species:** a nonnative species whose introduction, according to William Jefferson Clinton's Presidential Executive Order No. 13112, "does or is likely to cause economic or environmental harm or harm to human health"[16]
- **Native species:** with respect to a particular ecosystem, a species that has historically occurred or currently occurs in that ecosystem without human intervention
- **Nonnative (alien, exotic) species:** any species, including seeds, eggs, spores, or other biological material capable of propagating that species, that people have either intentionally or unintentionally introduced to a particular ecosystem
- **Species:** a group of organisms all of which have a high degree of physical and genetic similarity, generally interbreed only among themselves, and show persistent differences from members of allied groups of organisms
- **Noxious weed:** any plant (whether native or nonnative) designated by a federal, state, or county government as injurious to public health, agriculture, recreation, wildlife, or property
- **Weed:** any plant growing where it is not wanted and can be native or nonnative, invasive or noninvasive, and noxious or not noxious

Several of these definitions circle around two ideas: origin and harm. Invasive species must be both nonnative and harmful; they cannot be one or the other. They do not include native species that have suddenly proliferated and/or extended their range on their own. White-winged doves, for instance, are native to Texas, but they once were largely restricted to scrubland in the Rio Grande Valley. In the last few decades, because of habitat loss, heavy hunting in Mexico, and increased urbanization, they have shifted northward in Texas and beyond. No matter how much they proliferate in suburban settings, and even if their expansion may be influenced by human activity, they are not usually referred to as invasive in North America. In some circles the term "opportunistic" is used to describe the native white-winged dove's ability to enter and spread, usually taking advantage of some demographic or environmental change.

However, some consider the Eurasian collared dove, also increasingly prolific and spreading in Texas and elsewhere, as invasive, because the aggressive dove's homeland extends from Turkey to southern China. Humans transported the species into the Bahamas. From Nassau, escaped birds established a foothold in Florida (nesting in 1982) and spread west and into California in 2001, and Idaho in 2005, and now nest in at least three provinces in Canada. Curiously, it is yet to colonize New England.

People also transport native species to sites and areas outside their usual ranges both intentionally and accidentally. Intentional examples include the rainbow trout and house finch. The rainbow trout is usually found in the Pacific Northwest but has been released in rivers and lakes elsewhere in the

5500512

The Eurasian collared dove, abundant in Gulf Coast counties. Photo by Greg Bartman, US Department of Agriculture, APHIS Plant Protection and Quarantine program, Bugwood.org.

United States (as well as in South America and Australasia). The house finch originally resided in dry open country in the US Southwest and northern Mexico but was released in New York in 1940. It has flourished and spread in the east by settling in urban places and has also grown more migratory than in the west. The US tally of raspberry-jam-looking males and dull brown females is estimated to approach one billion individuals—though such numbers are hard to imagine. An example of an accidental spread of a native species is mesquite. Once restricted to western parts of Texas, the thorny, bean-producing tree has expanded across the entire state, sometimes forming dense thickets. People spread the tree through cattle drives (a single cow chip may contain 1,600 seeds), overgrazing on fenced land, and the suppression of prairie fires. In all these examples, we do not usually refer to the spread of these birds and trees as invasive because they were native to the state or country to begin with.

These extensions in a plant or animal's range raise an interesting issue about origin and political boundaries. Obviously natural ecosystems are oblivious to the lines we draw to divide counties, parishes, states, countries, and so on. In the previous examples— white-winged dove, trout, finch, and mesquite—the expansion of range happened within the same nation, so we tend to call these species native because they belonged to the region, even though following the definitions given earlier, they were not, strictly speaking, native to the new ecosystems they encountered. We tacitly say "close enough" and tend to reserve the term "nonnative" for imports and releases from regions from which organisms did not, and most likely could not, have arrived without human intervention or assistance.

A range expansion beyond national boundaries may be no different. For example, the nine-banded armadillo expanded its range into another country. Native primarily to Mexico, it did not head north into Texas until the latter half of the 1800s, though it now inhabits the entire southeastern United States. The armored mammal was not living here in 1492, nor likely in 1845 when the Republic of Texas entered the Union. If the odd-looking creature had crossed the Rio Grande in 1835, when Texas was still part of Mexico, there would be no discussion of its origin: one would simply say the native Mexican mammal was expanding its range northward. Only with the crossing of a political boundary do we start to quibble about origins. Interestingly, if we go back three million years, the armadillo's ancestors were not in what is now Mexico. Native to South America, they had only just scuttled across the Isthmus of Panama. To be really precise, we would define a species' origin by ecosystem and time, but in common parlance we rarely do this. Many of the worst Texas invasive species come from other continents (Asia, South America, Europe, etc.), so it is usually a moot point. In the case of the armadillo, the creature arrived here without human assistance, either by swimming across the Rio Grande or, as some claim, by walking beneath the water on the riverbed. So by definition the armadillo is now a native Texan species, which we have embraced as one of ours and declared our state's official small mammal.

Some experts have begun to question whether we are being overly pre-occupied with where a nonnative species originates or how far back we can trace it to label it native. In the United States, authors, planners, and others have often referred to a pre-1492 benchmark to describe what "pristine" North America looked like and whether an organism is indigenous. This 1492 geography, however, is more a metaphor than real. Pre-Columbian America was a settled and humanized continent (population estimates, however, vary hugely) well before Europeans arrived with their diseases against which native peoples had no immunity.[17]

We now understand that indigenous Americans had extensive trade networks and managed and tended landscapes (for example, selectively harvesting and spreading useful plants, setting fires to encourage open savannas and prairies) that colonists simply regarded as "wild," especially after plants and animals invaded areas after cultivators had died out. Compared with their own crowded, long-settled, and degraded homelands, colonists reportedly perceived North America as a "virgin" land full of native animal, plant, and timber resources untouched by humans and there for the taking. Using the 1492 benchmark for what is native serves to accentuate the mistaken belief that Columbus was the first person to introduce alien and mostly domesticated species. Domesticated animals and plants already existed and were being spread in the Americas prior to the arrival of Europeans. Two famous examples of nonnative domesticates are the dog and the bottle gourd, both of which are found in the New World between 9,500 and 10,000 years ago. Five hundred years before Columbus, a Viking foothold in Newfoundland is associated with exploration and the transshipment of resources, including timber and grapes, to Greenland. What the Norsemen carried into North America is subject to conjecture. They practiced a traditional North European agricultural lifestyle in Greenland. Whether they relied on planted grains or livestock from Greenland in setting up their camps on Vinland or elsewhere remains to be determined. About the same time, far away in the Pacific Ocean, Polynesians were feeding on sweet potatoes native to South

America. Switches were taking place within the New World and possibly far beyond—the Columbian Exchange simply accelerated them.[18]

Harm is also critical to the term "invasive." Thousands of nonnative organisms live within our borders, but they are not considered invasive unless they cause, or are likely to cause, explicit harm. Two nonnative trees, introduced at the same time and place, illustrate this point. Crape myrtle, the heat-tolerant, summer-blooming ornamental iconic to the US South, is native to China, Korea, Japan, and the Indian subcontinent. Homeowners and landscapers have extensively planted the shrublike tree since 1790 when French botanist André Michaux planted it in his garden near Charleston, South Carolina. But after 200 years of nurture that has produced close to 100 myrtle varieties and cultivars, there is little evidence, to date at least, that the crape myrtle is "established" or "naturalized" from growing self-sustaining populations. Part of the reason may be that many sterile hybrids are on the market. Saplings may spring up from a few roots in a house garden, but they do not do so in parks and woodlands or along roadsides, nor do they spring up among field crops. Crape myrtles, in other words, more or less remain where people plant them. They are "well behaved," as gardeners say. Purists may disapprove of these nonnatives and correctly argue that they take up space where a native shrub or tree belongs, which would be much more integrated into the ecosystem. In a minimal but still important sense, using time and energy on this substitute does harm; however, the myrtle does not disrupt ecosystems in the ways we insist invasive species do.

However, the chinaberry tree misbehaves, though we have been slow to recognize this. André Michaux also introduced this tree to his home near Charleston in 1790. For a while, this lilac-flowered, deciduous tree from India and tropical China was quite a hit, offering valuable shade along the streets of San Felipe de Austin, the unofficial capital of Stephen F. Austin's colony on the Brazos River, as early as 1828. By the 1850s, the dooryards of rural log houses as well as the more formal gardens of Texas' towns sported the fragrant "China trees." But unlike crape myrtle, chinaberry spreads readily by seed, with help from hungry birds. It started popping up on riverbanks and in abandoned fields, along roads and forest edges, and around older homesteads. By 1900 it was already well established in central and southern counties, and today the tree grows in every part of the state except the Panhandle. Recorded in 21 states, it is common in the South all the way from California to Virginia. All parts of the plant, but especially its fruits, are toxic to humans, cats, and dogs and somewhat to livestock (intoxications vary widely

Chinaberry tree in full fruit. Photo by Cheryl McCormick, University of Florida, Bugwood.org.

in severity and nature). The Texas Department of Agriculture added the chinaberry to its invasive plant list in 2013, recognizing the tree's potential harm and its resistance to insects and diseases. Chinaberry probably avoided recognition as an invasive plant because it was attractive, had a 200-year history in the state, and, importantly, has not been a problem in the state's agricultural areas, which is the reason it is still not on Texas' noxious species list and is still absent from those of many other southern states.[19]

The term "invasive species" can be misleading. The connotation of invasion is militaristic. It is as if a species, like a barbarian horde just over the hill and out of sight, were poised to fall upon unsuspecting citizens. It implies that an organism intends to harm people. But of course the animals and plants are simply doing what all species do, living and reproducing to the best of their abilities, and we are the "barbarians" for spreading them. Species that extend their ranges naturally are not considered invasive. We avoid using "invasive" in the main title of this book to prevent these deceptive overtones, yet we adopt "Unnatural Texas" with a good deal of irony. The animals and plants in this book are not artificial, and they are every bit as natural as any other organism on the planet. They simply have been transported to a place where they did not naturally (historically) occur and have managed to thrive on account of several shared characteristics (see table 1).

It turns out that only a handful of nonnative species actually get listed as

Table 1. Characteristics of invasive organisms

Fast growth and maturation
Rapid reproduction and large numbers of young
High ability to disperse, often over long distances
Plasticity to alter body size, shape, and/or behavior to adapt to conditions across a range of environments
A food generalist and opportunist
Association with humans, including desirability
Often earlier successful invasions (with natural enemies lacking)

Source: Ann K. Sakai, Fred W. Allendorf, Jodie S. Holt, David M. Lodge, Jane Molofsky, Kimberly A. With, Syndallas Baughman, et al., "The Population Biology of Invasive Species," *Annual Review of Ecology and Systematics 32 (2001): 305–32.*

invasive, but those that do are often very damaging. Current thinking, condensed into a statistical statement known as the "tens rule," holds that of every 10 species gaining access to a new area, only one will become established or naturalized; and out of every 10 established, only one will become invasive, that is, be classified as a weed or pest. At that point we condemn the organism as harmful because it hurts our pocketbooks, physical health, or the places where we live or play, perhaps combining all three. One should understand this tens rule as a rule of thumb somewhere between 5 and 20 percent. Ivan Jarić and Gorčin Cvijanović consider the tens rule to be an underestimate that masks higher numbers: "It might be more of an indicator of our lack of understanding of the impacts that introduced species, once established, produce," concluding that "the scientific community must be much more cautious and responsible regarding the message it delivers."[20] Others think the rule has worked quite well in explaining the numbers of high-impact nonnative species in the United States, including terrestrial vertebrates, insects, fishes, mollusks, and plant pathogens.[21]

When we focus on invasive organisms, we discover that for one reason or another native plants and animals are unable to compete with them and begin to decline or disappear. Foreign fish and amphibians, for example, directly prey on native species or, as do the zebra mussel, grass carp, Japanese kudzu, and red imported fire ant, alter or undermine the composition and functioning of natural ecosystems, aquatic or terrestrial, in which native species reside.

Popular interest and concern and our understanding about whether a living thing is nonnative or invasive are recent. In 2015, a global survey of the vascular plants across more than 80 percent of the earth's land surface discovered that 4 percent of them have been introduced and become naturalized in areas outside their places of origin. This may not sound impressive, but given the numbers involved, 4 percent comes to more than 13,000 plant species (the size of Europe's entire native flora). Experts are working to identify both established and newly arrived organisms, to partner with others, including nongovernmental organizations and the general public, to tease out lessons about dealing with invasive species. Their goal is to improve techniques for controlling these species on a real-time basis. To this end, North America is carrying the highest number of established nonnative vascular plants of any other major biogeographic region, including Europe. There is a clear need for coordinating efforts to assess which species are here and to monitor what is coming in or likely to arrive soon. This means communicating with neighboring states *and* nations, including source-region data sets that may indicate how best to deal with these species.[22]

The Eyes of Texas: Who Is Keeping Watch?

About 80 of the 4,800 or so higher plants in Texas are reportedly invasive. This percentage point for plants is lower than that for freshwater fish (3 percent) or for mammals (6 percent).[23] We want to keep percentages for invasive organisms as small as possible and have passed statutes and codes for agriculture, natural resources, and wildlife to demonstrate our commitment to controlling the arrival, establishment, and long-term impact of pests of one form or another (see appendix 1).

State lists name plants and animals that have inadvertently gained access to rivers, lakes, and marine and other habitats. Many of them have become problems among fishers, recreationists, water engineers, and city officials. Some of these naturalized Texans are targeted as economic pests and therefore deemed invasive; but others are tolerated or even grudgingly admired. Some are available for purchase in plant nurseries or as pets for the house or aquarium. Tucked in shipping containers, hidden in raw materials, or mixed in with manufactured products, new species are arriving by air, ship, road, and rail. However, more and more people are looking for them, especially when they know from past experience or from lessons in neighboring states or perhaps nations that the plants or animals are serious menaces.

Every day millions of pounds of vegetables, fruits, cut flowers, and other produce enter the United States, much of it passing through Texas via border crossings with Mexico. US Border Customs and Border Protection specialists check containers, trucks, pallets, and plants at ports of entry, including international mail points and air and sea passenger luggage terminals. Using sniffer dogs and machines, they look for the incidental entry of invasive species and for any deliberate acts of what they call agro-terrorism. In 2014, inspection specialists intercepted two insects unknown at that time in the United States. One was in a shipment of ginger examined in Pharr, and the other (during the same month of November) was discovered in a shipment of broccoli on the Rio Grande City International Bridge. Both shipments were refused entry and sent back to Mexico. Even the construction of new highways to old border crossings can raise concerns. Officials are anxious about the completion of Mexico's Federal Highway 40/40D, the so-called Inter-oceanic Highway connecting Mazatlán on the Pacific Coast to Reynosa, just south of McAllen, Texas, on the Gulf Coast, since the super-highway will make it that much easier for tropical pathogens and pests to reach our state.[24]

Earthquakes and weather events are natural or physical agents that disperse nonnative plants and animals. Five years after the 2011 tsunami associated with an offshore earthquake that destroyed the nuclear power plant in Fukushima and swamped other coastal communities in Japan, human-made materials encrusted with limpets, barnacles, sponges, sea stars, seaweeds, and mussels are still coming ashore along the Pacific Coast between Alaska and California. Transported by wind and waves, super rafts of floating plastic have carried almost 300 marine species, including fish, on all manner of objects consisting of children's toys to fishing boats, docks, and piers to seashore sites where specially garbed workers burn them with flamethrowers in an effort to keep newly arrived marine creatures from gaining a new start in the coastal waters of North America.[25]

Hurricanes rather than tsunamis devastate human communities in coastal areas and dislocate biological communities in Texas. These severe weather events may also serve as points of entry for nonnative organisms or assist the spread of ones already established. Reportedly, flooding associated with storms pushed the nutria's spread in both Louisiana and Texas and increased the dominance of Chinese tallow in hardwood forests around New Orleans. Fire ants can float for miles and retain buoyancy as a compact mass for several days before settling on drying earth and colonizing it. Storms damage and disturb existing landscapes and open them up for occupation by opportunistic and adaptable species, many of which are invasive (see table 1).

Ports on the Gulf of Mexico and cities on the rivers and water bodies that lead into the Gulf are also entry points for undesirable species. Houston, for example, is a major world-class port that welcomes 11,000 seagoing vessels every year and has received millions of metric tons of ballast water dumped by overseas vessels. The city's port, a 25-mile-long complex extending from close to downtown into Galveston Bay, is clearly a contact point for new and potentially invasive marine organisms. Ballast water has been implicated in the transmission of a cholera strain from South America to fish and shellfish in the Gulf of Mexico, as well as for a jellyfish from Australia that preys on other invertebrates in the same marine habitat. Regulations are now designed to reduce the advent of invasive species by treating ballast water, pumping it onshore to be disposed of, releasing it beyond US waters, or using water from an approved US public water system.[26]

Similarly, watershed networks and floods along river and lake edges also facilitate movement of alien plants and animals. In 1999, the Texas Parks and Wildlife Department (TPWD) published a 370-page book titled *A Guide to Identification of Harmful and Potentially Harmful Fishes, Shellfishes and Aquatic Plants Prohibited in Texas.* Parks and Wildlife personnel monitor aquatic species, most of which fall into families that contain 32 fish, 6 shellfish, and 18 aquatic plants. They also inform fishers and recreational boaters that it is illegal to possess, transport, or release any one of the listed organisms. It is easy for bits of lake or reservoir plants to stick to propellers or fishing gear and go on to pollute additional water bodies. Possession of the water hyacinth, giant salvinia, giant duckweed, water lettuce, hydrilla, and seven other plants is a class B misdemeanor under the Parks and Wildlife Code.

Scope and Structure

Apart from being some of the most problematic species in Texas, the animals and plants discussed here reflect the gamut of issues surrounding invaders. Some species demonstrate a classic version of the story: introduction, massive spread, obvious threat, immediate control. But in many ways, this trajectory is not as typical as we think. Many invasive stories are less straightforward, more nuanced, and open to question and opinion. Some species have become so common on our streets and in our yards that we do not regard them as exotic at all. Others did not prove to be invasive until decades after their arrival, and then only after we intentionally spread them or radically altered environments that gave them distinct advantages over native species. Our long and intimate relationship with certain species, such as pets and livestock, has allowed us to overlook the damage they do until recently, as we struggle (even morally) to come to terms with controlling them. Some of these species were imported to "improve nature" as much as to be specifically useful. Many were brought to our shores for the sheer pleasure they give us as adornments for a garden, pond, or aquarium. Others, highly endangered in their native lands, are imported to game ranches for hunting *and* preservation. Interestingly, we brought in most of these organisms intentionally; only a few snuck in by accident, but even these arrived because of the things we did.

All of these issues affect what we do about invasive organisms. Some species are so ubiquitous that eradication is out of the question, and we decide under what circumstances action must be taken. Others require constant control just to keep a situation manageable. Our love and fear of certain creatures bias efforts to control them. And the possibility that we could put invasive species to good use—harvesting them for food or oil—continues to lure us despite previous poor outcomes. Mechanical removal, hunting, and toxicants in gases and sprays have their pluses and minuses and often need to be applied in site-specific ways. Biocontrol, that is, using another (often nonnative) predator, parasite, or pathogen to attack, infest, infect, or otherwise control the invasive organism, is one of the latest and most popular tools in the defense arsenal and one that has shown good results. But bringing in an exotic to attack another exotic gives room for pause, no matter how effective it promises to be. Finally, the recognition that even invasive species offer ecosystem services reminds us to take care about removing them and in making sure that what they do in the larger biological community is being done or can be done by others.

The following chapters are arranged in a loose chronological order, not by the date of a species' introduction, or when it started to spread, but rather by when we began to *care*, which is to say, when we began to perceive that the species was causing harm. This arrangement, though imprecise and admittedly arbitrary, accentuates the cultural aspects of the invasive issue. Before the conclusion, we include a chapter highlighting some of the most recent introductions to the state.

1 Sparrows and Starlings
Dealing with Avian Reprobates

Until quite recently, transporting animals and plants around the world was accepted and openly encouraged. Nature enthusiasts thought the introduction of the English sparrow and common starling would be useful and actually "improve" birdlife in the United States and Texas. But their optimism quickly soured, creating the first beginnings of a sea change in attitudes and assumptions about importing new species.

Laura Joseph keeps tabs on her neighbors. She has recruited a cadre, who has taken to strolling across her large yard close to Austin's hallowed Barton Springs swimming pool. They are volunteers, who peer into the nest cavities of purple martins to note what is going on. Every year, as days lengthen and the weather warms, purple martins (*Progne subis*) return to Texas from Brazil's Amazon River valley, 4,000 or more miles away. They swoop down into more than a dozen or so houses and gourds that festoon Laura's overlook above Barton Springs Road. The bold purple males chuckle and gurgle as they claim nesting spaces, sparring with neighbors as they pick out and settle into cavities set on poles. Laura has more than 100 pairs of martins in her yard. Passersby admire their comings and goings and exuberant hubbub, while her dedicated martin-loving neighbors keep tabs on the colony and do what they can to help them nest successfully.

Martins are home birds. They gravitate to dwellings we construct for them, appearing to relish our companionship. Laura organizes helpers, who put out nest materials and eggshells for nutritious calcium. They record the eggs and chicks in each compartment during the breeding cycle. In Central Texas, purple martins nest from April through June, though Laura has recorded a pair or two even later. The kind of care lavished on this group of acrobatic and seemingly cheerful swallows draws praise from local media. Neighbors like to observe the birds and take pride in helping raise and fledge new martins before they head off toward the Gulf of Mexico and as far as Brazil, where most pass our coldest months. The most productive way to conserve this popular bird is to inspect nests regularly, which martins tolerate, and scare off or remove any and all predators that threaten breeding birds—especially the English or house sparrow and common starling.

Purple martin. Photo by J. J. Cadiz (CC BY 3.0).

Laura Joseph hosts a sizable purple martin colony in central Austin. (*inset*) Volunteers control invader starlings and sparrows and educate the next generation. Photos by Robin Doughty.

Both birds belong to the Old World. However, due to self-promotion, personal bravado, miscalculation, and culpable ignorance, both the sparrow and starling are abundant in the New World, much more so than the native purple martin. Both foreign species are common in Texas and hang about Laura's martin houses. Both have taken to neighborhoods, such as her spacious yard, and other areas, including city parks, lakeshores, and industrial areas. And both foreign bird species are invasive because they compete with native songbirds and damage crops. They also pose a special problem for purple martins because virtually all these summering swallows east of the Rocky Mountains nest in humanly provided nest spaces. Due to our prolonged interest and care for this largest US swallow and a dearth of suitable cavities in trees and buildings, such as store fronts (both also sought by newcomer sparrows and starlings), purple martins have come to rely on houses and gourds specifically constructed and raised for them. This is the chief reason that Laura and other martin landlords check birds regularly. They want to minimize the risks purple martins face from the two foreign interlopers.

Volunteers pull sparrow and starling nests out of martin houses. Elsewhere, martin landlords trap sparrows and starlings to keep them from raiding cavities and pecking the eggs and young of this graceful insect devourer.

House sparrow, an abundant and ubiquitous invasive bird. Photo by Vkulikov (CC BY-SA 3.0).

House Sparrow: Introduction and Spread

The so-called English sparrow (*Passer domesticus*) was the first of two alien songbirds to make a permanent home and become widespread in the United States, including Texas. The house sparrow, as people usually call it because of its habit of living in villages and towns, was deliberately shipped into this country. The director of the Brooklyn Institute, Nicolas Pike, reportedly imported eight pairs into New York in 1850. Caged and fed during the winter, the first batch of birds disappeared after release. Undeterred, while en route to a position as consul-general in Portugal in 1852, Pike went ashore in Liverpool, England. With $200 subscribed for another introduction, Pike arranged to send a shipment of wild sparrows and other songbirds to New York City. Caged on the steamship *Europa*, which docked on October 21 after a nine-day crossing, some sparrows found quarters in the tower of Green-Wood Cemetery chapel in Brooklyn. John Hooper, a member of the committee for house sparrow reintroduction, cared for them. In spring 1853, the cheeping birds found freedom on the grounds of Green-Wood Cemetery, which was the site for subsequent releases under the mistaken belief that additional British birds (including goldfinches, thrushes, larks, and nightingales) would spark new life among native avifauna.[1]

After this first successful release, hundreds of the brown-backed, gray-breasted, six-inch sparrows were shipped in from Europe, mostly from England and Germany, and enthusiasts freed them (and in time released their brethren hatched in the United States) in major cities across North America. Supporters loosed 12 birds in New York City's Madison Square in 1860, for example, and two years later added another bunch in Central Park, where passersby fed the familiar chirpers. Soon, Boston Common welcomed its first house sparrows, as did Quebec, Canada. Zealots introduced these likable birds across the Midwest and as far west as San Francisco and as far south as Galveston, Texas.

During the 20-year craze for the species, people liberated birds in at least 100 towns and cities, and more than a dozen of them, including Philadelphia, added an extra expense of importing sparrows directly from Europe. The general public swooned over "English" sparrows. "They are like gas in a town—a sign of progress," declared Colonel William Rhodes, having freed a number in Portland, Maine (in 1854), and then in his hometown of Quebec, Canada.[2]

Large bird colonies flourished east of the Mississippi River, with an outlier or two, one of which popped up in Texas, where J. M. Brown imported his sparrows directly from Liverpool to chirp and hop around his fashionable house in Galveston in 1867. A decade later, the English sparrow's range spanned more than one million square miles across 33 states and Canadian provinces. It remained a novelty in Texas. But Texans were well disposed toward it. In 1883, the general laws of Texas amended the Penal Code to make it a misdemeanor for anyone to kill or injure a sparrow, subject to a fine of no less than $5. The 1889 map of the English sparrow's range published in Walter Barrow's *Bulletin* shows a sizable distribution west of the Mississippi River into Missouri, with enclaves around New Orleans, Salt Lake City, and San Francisco. In Texas, colonies were yet to grow and disperse; only four Texas sites, including Galveston and Houston, are on the map. Residents were still getting to know the bird.

Changing Attitudes: The Lacey Act and "Rat of the Air"

Attitudes began to change. By 1900, disgust about the slovenly ways of this nonnative species and its aggressive habits toward native birds filtered into the halls of Congress. Senator John F. Lacey of Iowa condemned the English sparrow as "the 'rat of the air,' that vermin of the atmosphere," and wanted to prevent another "unbalance of nature" that he claimed it had caused, so he pushed his bill to keep sparrows out of the United States. It was far too late; however, the Lacey Act has helped conserve some animals at risk while also barring the importation of harmful ones. Under its first provision, the federal government sought to stamp out the trade in native game animals that may have been taken legally in one state and then shipped to another in which it was illegal to hunt them. Under its injurious import provision, the Lacey Act prohibited the introduction of four species and their eggs or offspring: the English sparrow, starling, mongoose, and fruit bat. The amendment to the Texas Penal Code was also rescinded.[3]

Since John Lacey's time the injurious species provision to his act has been amended several times and codified in the criminal code. In 1960, for example, an amendment to the list expanded to include fish, reptiles, and amphibians. At that time the sparrow and starling were dropped as injurious because "these birds have extended their range throughout the country and no feasible means of controlling their numbers or range has been devised," noted a Senate report.[4]

Initially, sparrow advocates reassured Americans that the foreign newcomers would pick off "inchworms" (actually caterpillars of a native moth) from springtime trees along city streets, avenues, and squares. These were insects that looped or "inched down" onto passersby and that native birds, frightened by the grime and clatter of industrial life, were disinclined or too wary to catch. Habituated to urban noise and hubbub, the house sparrow would deal with inchworms. "He is death to insects," declared Colonel Rhodes in 1877, even though "the bird is a little blackguard—fond of low society and full of fight, stealing and lovemaking." This sparrow had done well in Portland, Maine, and twice been secretly released by the colonel in Quebec, but had not survived. Then, reportedly in 1868, with publicity and fanfare, Colonel Rhodes's third attempt succeeded. Happily, Rhodes declared, "No live Yankee would wish to be now without the life and animation of the House Sparrow in his great cities." The Yorkshire-born colonel knew and loved his birds; two years after their success in Quebec, some sparrows appeared in Montreal, flourished, and spread.[5]

Not everyone was as bullish about English sparrows. By the time the last batches of the chubby brown chirpers fluttered out of shipping crates in the 1880s, bird scientists had judged it to have been a big mistake to have imported them. Bird-watchers condemned sparrows as "Bullying Britishers," encapsulating problems foreigners caused.[6]

A boorish manner and pecking grains added insult to injury. Sparrows failed to snap up street bugs, preferring seeds to caterpillars (they do feed some to their nestlings). Therefore, after scrabbling among fallen feed intended for city horses, street-urchin sparrows ducked into crevices in nearby brick walls and gutters. Loud chirps made by love-struck males multiplied in commercial and nearby residential neighborhoods. Bib-throated males were looking for mates and showing humans, sponsors or not, they were here to stay.

Sparrow War and Sparrow Pie

A "sparrow war" broke out. Supporters argued that sparrows were an industrious, well-known, brave, hardy, adaptable, friendly, and useful species. Similar anthropomorphic remarks by haters condemned them as dirty, oversexed, aggressive, urban immigrants that threatened the well-being and livelihoods of native birds. Rather than look at the sparrow from a biological perspective, commentators treated it as a tiny person who had ignored civic norms of respectability and behavior and was thus a bad example for teachers and children to witness. Sparrow haters also linked the bird to foreigners. After initially being welcomed as a nostalgic reminder of European homelands, this feathered immigrant was threatening to overwhelm American birdlife by sheer force of numbers, with scant respect for hardworking native-born American bird favorites, such as wrens, chickadees, bluebirds, and martins, which nature writers were encouraging the public to observe and study outdoors.

Others, however, had loftier goals for the English sparrow, at least initially. They envisaged a kind of marriage between natural history and literature, believing the sparrow was a sort of literary motif for people to appreciate. Mentioned a dozen times in the works of William Shakespeare, this instructional notion lent a cachet to the sparrow (and later to the starling, which appears only once in the bard's writings).

Including the house sparrows among additional birds, such as the virtuoso nightingale, bolstered the progressive impulse that Americans could and should "improve" the nation's avifauna. Reportedly the city sparrow, known and admired by Americans of European descent, was surely one such improvement because it was useful, willing to be a companion in bustling downtowns and neighborhoods, and easy to identify. Similar preferred species for American fields and gardens included European robins, common blackbirds, Eurasian skylarks, and various finches. The United States had beautiful-looking birds, but Europe had better songbirds, acclimationist improvers insisted: the sparrow fit somewhere in between; it was familiar, easy to get close to, and a useful inner-city bug eater.[7]

In fact, the English sparrow neither assisted farmers nor pleased or inspired city residents; rather, the opposite. Descriptions surfaced about sparrows chasing chickadees from feeding tables, displacing titmice and wrens from garden nest boxes, and attacking purple martins—an equally familiar but much more friendly and better-looking native swallow. The bullying and scruffy habits of sparrows infuriated enough people to cause them to characterize the sparrow's appearance—dumpy size, dingy color, squeaky voice, and dirty habits—as resembling that of a rat, not a bird.

Bird observers had noted that the sparrow-rats scrabbled after grain and picked over horse droppings in streets and gutters; they squeezed into pipes to produce young—brood followed brood as on a conveyor belt. These winged rats ousted native birds from nest sites and pecked at eggs and young. Refusing to eat many bugs like the neighborly martins did, the foreigners drove these native birds away, cramming nest cavities intended for martins with untidy straw. Rather than glide, circle, and warble above homes and gardens in search of harmful insects, the sparrows chirped incessantly and swarmed into fields to fill up on grain.

By 1890, matters had come to a head. Ornithologist Walter B. Barrows penned the very first bulletin of the USDA's new Division of Economic Ornithology and Mammalogy, titled *The English Sparrow in North America: Especially Its Relations to Agriculture*. Barrows's 405-page review and assessment indicted the English sparrow: "The Sparrow is decidedly injurious to grain, seeds of various kinds, and fruit; that it causes a decrease in the number of native birds in gardens and on farms, as well as in cities and towns; and that it is a serious nuisance in many ways."[8]

Keenly interested in how birds could be helpful or harmful to farmers, Barrows, who worked in the Division of Ornithology and Mammalogy (later the Bureau of Biological Survey) in the USDA from 1886 to 1894, included ample expert and firsthand testimony about birds pecking grapes, orchard blossoms, fruits, and vegetables. They ate oats, corn, rye, wheat, barley, sorghum, and rice. In respect to aggression toward native birds, Barrows, who wanted to assess the status of introduced species, reported 70 kinds of native birds molested by English sparrows. He listed 377 reports of injury to the eastern bluebird, 263 to the purple martin or "martins" (it increases to 347 if "swallows" are included), 189 to native sparrows, 182 to American robins, and 180 to wrens. He even reported sparrows abusing the feisty northern mockingbird and aggressive blue jay.[9]

The eastern bluebird, "one of the pluckiest of our native birds," noted Barrows, cannot hold its own against the sparrow. The purple martin did better than similar nest-box species because it was larger than the sparrow and nested colonially, ganging up on the invasive birds (but not consistently). Barrows included one report from Kansas that the sparrow "pulls the Martin and Swallow from their nests and throws out the eggs." Another from Illinois commended the martin for resisting sparrows, only to ultimately fail to protect its brood while away catching insects. In Elida, Allen County, Ohio, S. D. Crites reflected dolefully, "I have watched the battles between Sparrows and Martins, by the hour. Now [in 1886] there is not a Martin to be seen in the country." The sparrow was relentless, Barrows argued, contradicting comments about the bird's tolerance of native species by including reports about it wearing down purple martins through its tenacity and "sheer force of numbers." Evidence he cited came from about 110 replies to a questionnaire sent out to members by the American Ornithologists' Union in 1883, which was then greatly augmented by more than 3,000 replies to a circular distributed by the USDA in 1886. Commonly, reports referred to the English sparrow in military terms. People characterized it as the purple martin's "bitter" enemy, of surrounding a colony, then sending in squads to route martin defenders.[10]

Barrows concluded, "The Sparrow does no kind of beneficial work as an insect destroyer which would not be much better done by native birds; while its presence prevents other birds from accomplishing many kinds of work which the Sparrow does not undertake at all." The bird promoters had got it all wrong; the English sparrow had done no good at all; in fact, it was doing plenty of harm. After he moved to East Lansing, Barrows continued to work on food habits of birds as professor of zoology and physiology at Michigan Agricultural College (now Michigan State University) until his death from apoplexy in 1923.[11]

Other bird experts applauded his review and assessment and rallied around Barrows's official condemnation. He had in fact confirmed what the American Ornithologists' Union had concluded five years earlier in an 1885 judgment of the "Sparrow Question." Opinion ran strongly against the species, and an ornithologists' committee concluded it had been a mistake to introduce it. The public was weary of sparrows.[12]

People had begun to suggest, as Barrows did, that farmers lace grains with strychnine or arsenic and poison the foreign birds. Others took a different path. In place of using pigeons, some entrepreneurs had begun to trap the English birds and use them as targets for "sparrow shoots." W. T. Hill of Indianapolis, for example, trapped at least 40,000 sparrows over a two-year period ending in October 1887 and still considered them to be "superabundant."[13]

Gourmet-minded folk touted sparrows as tasty morsels in pot pies. Citing American publications, Barrows noted that on July 20, 1887, the *New York Times* declared, "Sparrow pie is a delicacy fit to set before a king," and notified readers that a German woman on Third Avenue was trapping 75 sparrows per week. She cooked them for boarders, running a little wooden skewer through every carcass. She preferred basting the choicer meat of the female birds.[14] In November 1887, a well-known game dealer in Albany, New York, purchased 3,000 sparrows, paying $1 per 100. Youths immediately killed a lot more.[15]

Continued Expansion

In Texas, in addition to sparrows being shipped into Galveston and released there on three separate occasions, the nonnative bird appears to have ridden hobolike in boxcars from one town to another. Some said they roosted in boxcars and fluttered out at each stop. Others speculated birds swooped to follow seeds and grain spilled along the tracks. Whatever the truth, the English sparrow turned up in Austin in 1895, reached San Antonio a year later, and reached Brownsville in 1905. There are no records of deliberate releases in any of these cities. By that time the public was sick of the bird but unable to prevent its spread.

Today, the invasive house sparrow is a permanent resident in all 254 counties. It appeared in Langtry in about 1901 and flourished as far west as

Marfa, where a hailstorm in 1923 reportedly killed several hundred.[16] Individual sparrows have survived as far north as Anchorage, Alaska; and in 2007 five birds turned up in Alaska's Bering Strait National Reserve. This surprise arrival sparked speculation whether the sparrows had flown in from Russia rather than from the United States.[17]

The house sparrow resides in all US states and has pushed south into Central America as far as Panama. It has also fluttered geographically across Argentina, where misguided enthusiasts released some birds in 1873. By the turn of the last century, sparrows lived in Brazil and Chile and by the 1920s were eagerly chirping on the Falkland Islands. Additional releases in southern Africa, Australia, and New Zealand (from where people carried it to Hawaii) contribute to the species' reputation as the most widely distributed bird in the world.[18]

However, since the 1970s, house sparrow populations have mysteriously and significantly declined across Europe, particularly in its UK stronghold. The sparrow has disappeared from many areas of London, for example, where it was once common. Researchers are puzzled by this collapse and have tentatively concluded that insufficient food for nestling sparrows causes many to starve.

The European or common starling. Photo by Pierre Selim (CC BY-SA 3.0).

Many martin landlords use "excluder" devices to prevent starlings from entering nest cavities. Photo by Robin Doughty.

The European Starling

The European or common starling (*Sturnus vulgaris*) appears on a list of desirable new species drawn up by bird lover Eugene Schieffelin, who also liked sparrows, of New York City. Reportedly, Schieffelin, a wealthy drug manufacturer, released 100 European starlings in two batches in 1890 (60 birds) and 1891 (40 birds) into New York's Central Park. This was 40 years after he turned loose 12 house sparrows on Madison Avenue where he lived. Dutifully, the starling repeated sparrow antics, expanded its range, and exploded in numbers. Today, experts say 200 million starlings inhabit North America—that is, about one-third of all the world's starlings.[19]

Like the house sparrow, the starling has acquired a reputation of being aggressive and even predatory, and both farmers and ranchers blame it for damaging crops and spreading disease. The USDA has documented starlings pilfering and pecking into berries, figs, and grapes; gobbling grain; contaminating cattle feedlots with *Salmonella*; and spreading other diseases. Starlings and sparrows rank second to the feral pigeon in the number of requests US states make for federal damage control and assistance. Flocks of resident and migratory starlings often swarm around feedlots in Texas, for example, and roost on downtown buildings, in city parks, and in woods in large numbers. Droppings of millions of starlings can eat into brickwork, kill trees and shrubs, and pose risks to human health. One estimate puts the damage done by starlings in the United States at $1 billion annually.[20]

Threats to Aircraft

The same flocking habit that birds adopt as they forage in landfills and garbage dumps has also downed commercial aircraft. Low-value sites, such as landfills, often close to airport runways make aircraft subject to so-called bird strikes. On landing and takeoff, an airplane may lose power suddenly after smashing into scores or even hundreds of eight-inch-long starlings. This happened with tragic consequences at Boston's Logan airport. On October 4, 1960, a Lockheed Electra turboprop bound for Philadelphia was cleared for takeoff at 5:39 p.m. A few seconds after becoming airborne, the airplane struck hundreds of starlings that had erupted from a nearby landfill into its flight path. The engines of the Eastern Airways 375 flamed out. Within one minute after takeoff the flight crashed into Winthrop Harbor at the end of the runway and broke into two pieces, ejecting eight passengers and two flight attendants. Sixty-two additional passengers and crew died in the accident; only one survivor escaped without serious injury.[21]

Bird strikes caused by starlings and blackbirds (the species are often lumped together) have increased over the years, topping 1,700 reports between 1990 and 2001. Impacts can be catastrophic, as jet engines literally ingest birds. Estimates of damage and flight delays due to bird strikes to commercial aircraft worldwide cost a minimum of $1.2 billion annually. However, most strikes, possibly as many as 80 percent, are not reported, and the ones that are usually minimize damage.

Although starlings are smaller and have just a fraction of the weight of ducks, geese, swans, large hawks, and vultures, which present serious hazards to passing aircraft, they do pose a serious risk. A large, compact flock hit an airplane in Dusseldorf, Germany, in September 2009. On takeoff, Germania B737 ingested starlings into its right engine and was forced to return to the airport. That month, 14 additional known bird strikes occurred from places as far apart as China, Turkey, India, Brazil, and the United States. The loss of 190 people and 52 civilian aircraft had been attributed to bird strikes up to that time.[22]

Spreading like Oil Smoke

The starling is more migratory than the house sparrow, and it spread across the United States without human assistance. This ability to disperse and migrate in flocks over long distances has contributed to its status as a common species in North America. Researchers have banded starlings and noted that some individuals may cover 300 or 400 miles between capture sites and feeding areas, such as the feedlots in Texas.

Dispersal is an ingrained habit. In Europe, millions of starlings cross the North Sea from continental Europe for the winter. During the day, birds forage in fields and pastures and then congregate in huge flocks that swirl, dip, and dive over roost sites, often selecting reed beds as dormitories. Through media coverage, thousands of people gather to witness this wildlife spectacle. It happens over well-known marshlands, for example, close to Glastonbury in Somerset and near Otmoor north of the city of Oxford.

The same roost behavior exists in the United States, where during colder months starlings and native blackbirds come together. However, rather than seek out these sites as spectacles, the public avoids them. The accumulation of droppings from millions of roosting birds kills off vegetation, smears trees, and harbors transmissible diseases. Omaha, Nebraska, was faced with 25,000 roosting birds and resorted to noisemakers, repellents, and poisons to clear them from downtown. Feedlots also suffer loss and contamination from starlings; so do grain growers and horticulturists. At least 20 bacterial and fungal diseases are associated with starlings, including salmonellosis, vibriosis, avian TB, and histoplasmosis.

The USDA's Wildlife Services, which addresses complaints about animal damage, notes that the common starling ranks second to the feral pigeon

A murmuration of wintering starlings in southern England. These undulating masses attract tourists but do pose hazards to aircraft in certain locations. The birds' droppings are messy and harbor diseases. Photo by Tanya Hart (CC BY-SA 2.0).

among the 97 established nonnative bird species in the United States. It is an invasive species in terms of crop damage, disease transmission, and competition and predation among native birds. Wildlife Services estimates crop damage in excess of $150 million annually.[23]

Harry Oberholser's *The Bird Life of Texas* notes that starlings first appeared in Texas 25 years after being released in New York City. A single bird was picked up dead in Cove, Chambers County, in December 1925, and a second one taken in Beaumont, Jefferson County, in early January 1926. Both starlings came from English stock that Schieffelin and his Acclimatization Society friends had released in New York City. Previous efforts in the 1870s to introduce birds into Cincinnati, Ohio, and Quebec, Canada, had failed.

However, not long after the first starlings arrived in Southeast Texas, others turned up as far south as Port Isabel. Oberholser, a respected ornithologist, who worked for the US Bureau of Biological Survey (later the Fish and Wildlife Service), recorded a remarkable influx of birds in the winter of 1933–34. Starling flocks resembled a "thick pall of oil smoke" over San Antonio, Oberholser noted. Two years later, a single starling had reached El Paso, and a decade later other birds had flown into California.[24]

It took longer in Texas for wintering starlings to stay and breed. However, by 1974, when Oberholser's classic bird book appeared, starlings nested from sea level to 3,600 feet; only the desert area beyond the Pecos River was free of them. Elsewhere, starlings had been recorded in all lower 48 states and were beginning to turn up in Alaska. Today, the species has made its home in desert areas too, including West Texas and northern Mexico, although numbers remain low. However, data suggest that the overall population of the common starling in Texas has doubled over the past 50 years.[25]

The Purple Martin's Nemesis

The aggressive starling from Europe carries a menacingly sharp yellow beak that it jabs at other species, including woodpeckers, in order to secure nest sites in trees, birdhouses, utility poles, and even traffic signs. The bird is very adaptable, feeding on grains, insects, and household refuse. It is unafraid of people and does well in metropolitan areas. Declaring that "starlings are capable of seriously reducing martin populations," bird specialist Charles R. Brown referred to a critical ecological issue posed by this invasive bird. Brown has appealed to martin "landlords" not to allow alien starlings to use martin houses.[26]

Initially, the damage caused to purple martins seemed secondary to that caused by the house sparrow. Both birds have been and remain aggressive toward native bird species. The onus for exclusion has shifted more and more toward the starling, because this black bird with a yellow, speared beak enters purple martin houses and not merely evicts the tenants, as many sparrows do, but murders adults and young and destroys eggs. Jim Gardner, a close observer of the martin colony near Barton Springs in Austin, states that the starling is a real killer of martins and has seen what they have done.

Competition and predation are pronounced in the purple martin's stronghold states, such as Texas, but happens across its range. Martin land-lords remind us it is better *not* to put up a martin house, or locate one close to trees or in similar unsuitable sites, because martins will not use it, but starlings will. Often, experienced martin caretakers take a hands-on approach, which means pulling out the nests of starlings or trapping and removing both starlings and sparrows.

Recently, martin lovers have adopted variously shaped entry holes that allow a purple martin to pass into the nest cavity but prevent European star-lings from gaining access. This Starling Resistant Entrance Hole (SREH) is a crescent-shaped entrance (not the conventional oval one) that makes it diffi-cult for the longer-legged starling to squeeze through.[27] A newer bat-winged design, invented by Duke Snyder in 1999, also excludes starlings. However, house sparrows can negotiate entryways that exclude starlings.[28]

Current opinions and suggestions on websites about martin biology and conservation represent decades of trial and error. One hundred years or more ago, a pioneering generation of bird lovers experimented by pro-viding ample housing for all three birds, hoping that sparrows, starlings, and martins could live together. They were disappointed. The two foreign bird species jostled and disrupted martin reproduction. Sparrows and star-lings entered cavities already claimed by martins, even though empty ones remained nearby. These experiments about coexistence failed; negative regard for starlings and sparrows has continued.

In fact, if anything, opinion has hardened in recent decades. Bird scien-tists blame declines in purple martin populations in the Great Lakes states, New England, and the Pacific Northwest to an increase and spread of Euro-pean starlings. The western martin, a recognized subspecies of the purple martin, which nests in California, Oregon, Washington, and as far north as Canada's British Columbia, has traditionally sought out natural cavities in which to nest. These paler birds use abandoned woodpecker holes and nat-ural snags in dead trees or tree limbs rather than humanly provided cavities. Starlings, however, have commandeered these holes and made it harder and harder for native martins to find suitable nest sites.

Volunteers have been heartened to note, around Puget Sound and farther north along Vancouver Island, for example, that purple martins have started to breed in single nest boxes nailed to pilings in coastal bays and sounds. Starlings do not frequent these sites. Martin numbers have actually increased around Puget Sound, where they once were common until starlings arrived, notably near Seattle. Purple martins are also responding well to human-made structures in British Columbia, Canada, where experts select sites away from the croplands and orchards preferred by starlings and sparrows.

The Lacey Act sent a strong signal about our attitudes toward wild animals. It exemplified a shift from unqualified enthusiasm for importing new species, typified by the widespread and prolonged introduction of house sparrows and starlings, to concern and disapproval about the injurious impacts these and other newcomer animals were having on the land and life of the nation. The federal law listed injurious animals, which were prohibited from entering or being shipped across the nation. The sparrow and starling were catalysts for this federal mandate, and more than 200 additional animals have joined them since its inception. The die was cast in 1900: some animals (native species) were good; other animals (notably foreign species) were bad—the English sparrow and European starling demonstrated this fact. Accurate information, careful observation, and expert analysis led to concerns about other nonnative species. It was the lesson Lacey and his colleagues intended to share so that injury from foreign mammals and birds (and later for plants) could be anticipated and thus prevented.

2   Water Hyacinth and Hydrilla
The Dastardly Duo

Within a decade of its introduction, the ornamental water hyacinth posed such a serious threat that the US Congress passed laws and appropriated funds to control it and debated whether importing a plant-eating mammal from Africa would keep it in check. The recent aquatic invasion of hydrilla demonstrates that some alien plants add severe economic and environmental burdens to rivers and lakes. And by altering watersheds and building reservoirs, we have inadvertently favored the establishment and expansion of seemingly innocuous plants (and animals). Having been discarded from aquaria or concealed in boating equipment, these often-inconspicuous organisms take hold in artificial water bodies. Both water hyacinth and hydrilla have been managed with biocontrol agents, never adopted for sparrows or starlings, and both aquatic invaders can be regarded as beneficial in certain circumstances, complicating our strategies about dealing with them.

A blooming mat of water hyacinth on Lake Conroe, Texas. The striking flowers played a major role in the plant's US introduction and initial spread among water gardeners. Photo by Roy Luck (CC BY 2.0).

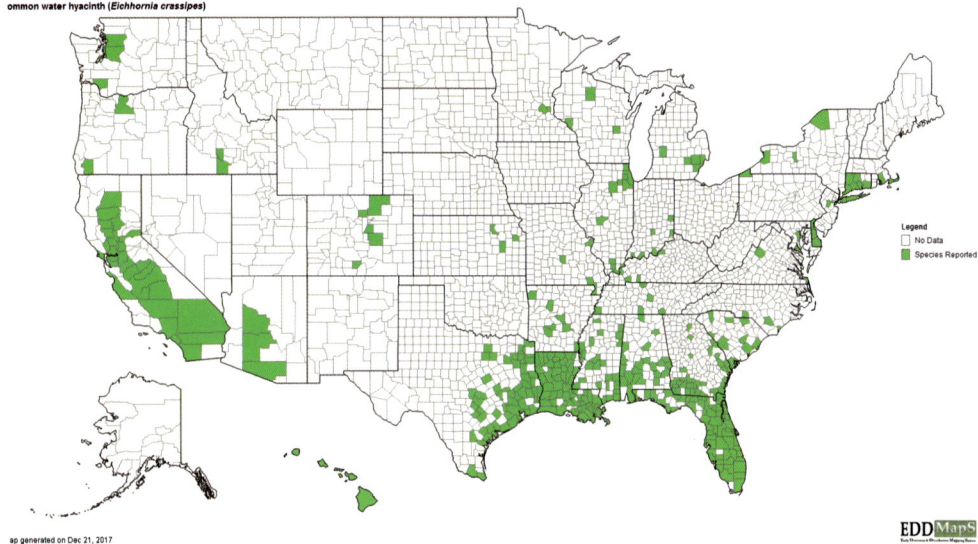

US distribution of water hyacinth (*Eichhornia crassipes*). Source: EDDMapS, University of Georgia—Center for Invasive Species and Ecosystem Health, 2017, http://www.eddmaps.org/.

At the turn of the twentieth century, the US southern lowlands were in trouble. A floating exotic weed, introduced as a pond garden ornamental a dozen years earlier, had grown so explosively in the bayous and rivers along the coast from Florida to Texas that shipping routes were becoming unnavigable. Giant floating mats of the once beguiling water hyacinth (*Eichhornia crassipes*) were impeding and even halting the shipment of millions of tons of freight. One engineer noted, "It matters not how wide your stream is, or how deep, it will not be navigable throughout the year if the hyacinths are allowed to take possession of it."[1] A canal needed to be exposed to the non-native plant only for a year to two before it became impassable. Accumulating in narrowing channels after floods, masses of water hyacinth had even overturned bridges.[2]

No one defended the lovely weed. No "hyacinth war" broke out between supporters and detractors, as it did with sparrows. This time militancy came from the top, from lawmakers who targeted the foreign invader plant. With the passage of the Sundry Civil Appropriations Act (June 4, 1897), Congress recognized the water hyacinth as a serious problem and authorized the secretary of war to investigate the plant's obstruction of ship traffic in the South Atlantic and Gulf States. The US Army Corps of Engineers, born in 1802, has battled the plant ever since.[3]

Enlist the Hippos?

Around 1900, America was suffering a serious meat shortage. Like the hyacinth, city populations were exploding as a result of prolonged immigration. Earlier, westward expansion had provided a safety valve to food scarcity, but overgrazing by livestock had damaged rangelands, and the meat industry—which Upton Sinclair famously took to task in his novel *The Jungle* (1906)—could not keep up with demand. Delmonico's, the famous New York steakhouse, was raising prices on everything. People even whispered about eating dogs.[4]

A Louisiana congressman proposed a strategic response to both problems: import the common hippopotamus from Africa. In what seems today a joke, a bill—the American Hippo Bill (H.R. 23621)—put forth in Congress in 1910, aimed to import and release hippos into Gulf Coast swamplands. Introduced by Representative Robert "Cousin Bob" Broussard from New Iberia and championed by Theodore Roosevelt, the huge aquatic herbivores would munch on hyacinth and then we would munch on them. The *New York Times* supported the proposal as "practical and timely" and touted hippopotamus brisket, which is actually quite tasty, as "lake cow bacon." One way of dealing with degraded rangelands was for southerners to become hippo ranchers and supply the nation with choice fresh meat. William N. Irwin, a (misinformed) researcher of the Bureau of Plant Industry at the USDA, told the *Washington Post*, "I hope to live long enough to see herds of these broad-backed beasts wallowing in the Southern marshes and rivers, fattening on the millions of tons of food which awaits their arrival."[5] Irwin even proposed *seeding* water hyacinth across 10,000 square miles of marshland in hopes of producing an enormous quantity of meat valued at $100 million at the time.[6]

In all this fervor no one seemed to have considered that the 3,000-pound pachyderms are unmanageable and regarded as one of the most dangerous animals in Africa. And no one bothered to check that they graze almost exclusively on the grasses around their river and lake homes (venturing up to two miles from water at night), and they usually avoid aquatic plants. The hippo bill created excitement but fortunately never came to pass.[7]

The "World's Most Troublesome Weed"

By most accounts, the water hyacinth made its American debut as a beautiful, seemingly harmless souvenir handed out by Japanese government representatives at the World's Fair in New Orleans in 1884, also celebrated as the World Cotton Centennial. Sporting gorgeous spikes of lavender-pink flowers flecked with yellow spots, the plants grew well in park and garden ponds wherever winters were mild. They were sure to thrive in the water gardens in cities such as New Orleans.[8]

Proliferating in surrounding parishes, within a few years the pretty hyacinth insinuated itself into the lakes, streams, and watercourses that make up the Mississippi delta. By 1890, the same year that Eugene Schieffelin began releasing starlings in New York City, the hyacinth had already popped up in Florida, where within five years giant floating mats of the weed, as large as 75 yards wide, stretched along the St. John's River for more than 100 miles.

Water hyacinth spread quickly beyond Louisiana and Florida. By 1902 Congress had begun authorizing funds to remove the weed from navigable waters in Texas. But it was essentially too late; the plant was growing in the Port Arthur Canal, a lumber- and grain-hauling waterway between Taylor Bayou and Sabine Pass, and was snaking up the Sabine River as far as Orange.[9] Water hyacinth soon became naturalized in all southeastern coastal states and up the Atlantic Ocean seaboard as far north as Virginia. It made its way into California. The plant also grows, often densely, in Arizona, Kentucky, Tennessee, Missouri, and New York, as well as in Hawaii, Puerto Rico, and the Virgin Islands. States with particularly bad hyacinth infestations include Louisiana, Florida, California, Hawaii, and Texas. The plant is now common in the Lone Star State from the Louisiana border south to Brownsville and infests

Water hyacinth crowding Lake Raven in Huntsville State Park, Texas. Photo by Roy Luck (CC BY 2.0).

35 inland water bodies covering a total of 10,000 acres. It has been a particular nuisance in Caddo Lake, where it showed up in the state's only natural lake around 1950, and in Toledo Bend, Texas' largest reservoir; and it heavily infests wetlands in southeastern counties.[10]

The weedy explosion in the United States is not unique. Water hyacinth has spread to more than 50 countries on all continents, except Antarctica, and appears on the International Union for Conservation of Nature's (IUCN) list of the 100 most dangerous invasive species. It is vilified as "the most renowned of all aquatic weeds,"[11] "the world's most noxious invader of wetland environments during the past 150 years,"[12] or simply "the most troublesome weed of the world."[13] In fact, even the souvenirs distributed in New Orleans did not come from the plant's original homeland but from the Orinoco River in Venezuela, where the water hyacinth had invaded and was a "most serious pest."[14]

Water hyacinth's original home is Brazil. It is indigenous to the Amazon basin and the freshwater lakes and marshes farther south in the Pantanal, the world's largest seasonally flooded wetland. The Pantanal is 10 times larger than Florida's Everglades and is shared with Bolivia and Paraguay as well as Brazil. There it grows in still or slowly moving water, preferring warm temperatures (82°F–86°F) and high light intensities. It does not like freezing temperatures but can withstand light frosts.

If this sounds like a good fit for the entire southeastern United States, if not most subtropical waterways around the world, it clearly is. Interestingly, recent DNA analysis reveals that about 80 percent of adventive water hyacinth populations—that is, plants outside their native range—are genetically uniform. They belong to a clone that accounts for nearly 80 percent of all individuals sampled in the area they have invaded. Although the number of separate introductions is not known, it is clear that a very few hyacinths have gone a very long way.[15]

Biology and Growth: Floating Mats

Water hyacinth is an erect, free-floating, perennial aquatic herb. A single plant is usually composed of six to eight leaves, radiating in a rosette from a central point. The leaves may rise up to five feet above the water's surface. Their inflated and spongy stalks, composed of air-filled tissue, buoy up the hyacinth so that in windy conditions, it sails along, even against a current.

Hyacinth flowers do set fruit, though these are rarely visible because the flower stalks bend below the water to release more than 3,000 seeds, which sink into bottom sediments where they remain viable for 15 to 20 years. Seeds

A young water hyacinth showing the inflated stalks that keep the plant afloat, the dense roots, and the runner that gives rise to, and maintains connection with, its offspring plantlets. Photo by Forest and Kim Starr (CC BY 2.0).

are not, however, a major means of reproduction in most habitats where constant-level water and the dense shade from parents floating on the surface prevent exposure and the quantity of sunlight needed to germinate.

Water hyacinth's main reproductive strategy is asexual or vegetative, which means the plant clones itself by creating new plantlets along shoots. It produces so many runners and daughter plants that clusters of 25 to 30 interconnected plants are common. The runners, together with the hyacinth's thick clusters of profusely branched roots hanging in the water, form densely interwoven colonies. In ponds and lakes and in places with very slow currents, such as drainage ditches, canals, and bayous, hyacinth colonies form continuous mats of outward- and downward-growing biomass, part living, part decaying. These may grow more than six feet thick and touch the bottom of a water body.

In essence, these mats are floating platforms. A medium-sized one, containing approximately 800,000 plants per acre, has a loading capacity of nine pounds per square foot in the summer months. Local citizens claim a mat can easily support the weight of a dog or child and, when two feet thick or more, bear the weight of a man. In one case a cow was reported to be grazing contentedly on top of a mat. A Louisiana engineer noted a boat with a three-foot draft grounded on a mass of hyacinths in a bayou 25 feet deep.[16]

Over time, especially in calm and stagnant conditions, these floating mats stabilize and appear permanent. Constant growth builds up a soggy debris layer, which invites colonization from other plants; ooze also builds up on a lake- or streambed below. These platforms, called floating islands or flotants, accelerate ecological change. Additional water plants like broad-leaf cattail, maidencane, and bulltongue arrowhead colonize them and may eventually crowd out the hyacinth.

One 10-year study found that 33 aquatic, 21 wetland, and 9 terrestrial plant species had become established on hyacinth platforms. As the platforms thicken, they resemble land and sharply reduce areas of open water, converting lakes and ponds into marshes. One researcher speculated that the rapid spread of hyacinth across Louisiana's marshlands brought about more ecological change (i.e., the disappearance of open water) in the 50 years after 1890 than in the preceding 2,000 years.[17]

As these platforms demonstrate, water hyacinth grows prolifically and invades rapidly. Plants may double in number every two weeks. A bayou with only a narrow fringe of hyacinths along its shore in March will be completely covered in June. Twenty-five plants can generate enough biomass to cover 2.5 acres of water area and weigh as much as a fully loaded jumbo jet—all in just one growing season.[18]

Threats of the "Giant Sponge"

Water hyacinth turns a thriving freshwater community—something mostly liquid—into a hulking mass. This mass clogs pipes, intake pumps, and hydroelectric and irrigation systems. Hyacinth "infestations can render multi-million dollar flood control and water supply projects useless," note state biocontrol researchers.[19] Acting as a giant sponge, they impede water flow through canals and back up runoff to cause flooding.

Water hyacinth colonies also transpire water. Estimates of water loss are as much as 13 times higher from plant colonies than evaporation from open water. Researchers in Africa, for example, have made the astounding claim that evapotranspiration from hyacinths in Lake Victoria may reduce the flow of the world's longest river, the Nile, by as much as 10 percent.[20] That would be about 68,000 gallons per second. Put another way, in Texas, water pumped into the atmosphere by hyacinths may total 100,000 acre-feet per year, enough water to supply more than a million people.[21]

Ecological damage is less apparent but equally critical. Hyacinth mats limit gas exchange between the atmosphere and the water, block sunlight from reaching phytoplankton and underwater plants, and create a constant supply of decaying matter and detritus. Such conditions reduce levels of dissolved oxygen, increase carbon dioxide concentration, and acidify the water. In fact, the oxygen-depleting pollutional load of a one-acre mat reportedly equals the sewage created by 40 people.[22] The reduction in phytoplankton weakens invertebrate communities and ultimately drives away the fish that depend on them. The floating mats tear up submerged and floating plants, outcompete native plants, displace forage used by wildlife, and alter the entire aquatic habitat.

The ensuing conditions also increase habitat for mosquitoes and other disease vectors by making it difficult for natural predators—or insecticide applications—to reach them. Malaria, encephalitis, and schistosomiasis were until recently the main human diseases of concern, but global warming and the introduction of new diseases have expanded the list to include West Nile virus, dengue fever, chikungunya, and, most recently, Zika virus.

Methods of Control: Pulling or Spraying

In the early 1890s, faced with the daunting prospect of more and more hyacinths, sawmill workers along Bayou Plaquemine, which drains into the Atchafalaya River in Louisiana, simply pitchforked the plants onto its banks. In 1899, backed by $25,000 appropriated by Congress, a stern-wheel steamboat mechanized the process. Known as the "Lily Boat," the *Ramos* was rigged with a conveyor belt that pulled plants out of the water, crushed them in a set of rollers, and deposited the pulp on the bank, a process modeled after Louisiana sugarcane processing. It removed about 188,800 square yards of plants before the funds were exhausted.[23]

The bottom line, then and now, is that mechanical removal is both slow and expensive. Even in 1900 the process cost six cents per square yard of hyacinth removal. Within two years the *Ramos* was rerigged for spraying a patented chemical on floating hyacinths. Between September 1902 and May 1903, it sprayed 523,000 gallons over 3.5 million square yards of plants and practically cleared Bayou Plaquemine. The next year, the *Ramos* cleared eight bayous and rivers and a watchman was positioned on the Bayou Plaquemine boom to check that passing vessels were free of hyacinths. Spraying continued the following year under a $40,000 appropriation. With the purchase of a second steamboat, aptly name *Hyacinth*, operations began in the tributaries of the Sabine River and the Neches River in Texas. The chemical proved 24 times cheaper and much easier than relying on the original mechanical belt.[24]

Similar removal techniques, however, do exist elsewhere. In Africa, the mechanical removal of hyacinth from Lake Victoria costs $6 million to $20 million annually and just barely keeps the weed in check. Aside from the cost, only one or two acres of the monster mats can be removed daily; the plant simply grows more rapidly than it can be eliminated. And there is always the problem that machines break up the plants into bits that can lead to further spread. In short, mechanical removal is at best a short-term solution to a long-term problem.[25]

Chemical control, while more cost effective and less labor intensive, is always problematic. The captain of one of the early spray boats, using arsenic, died from poisoning and 12 of his crew were sickened. While herbicides are now selected with care, there are always fears that there will be unintended consequences—to water quality, fish stocks, microalgae, and the food chain; or that they will seep into the groundwater and affect drinking supplies. Chemical control is best used in small areas that are severely affected.[26]

Three common herbicides used to attack hyacinth are diquat, glyphosate (marketed by Monsanto as Roundup), and 2,4-D, one of the two main ingredients of the defoliant Agent Orange used in the Vietnam War. The problem is that herbicides kill the prominent, exposed plants but leave hidden ones, fragments, and seeds untreated, allowing reinfestation. Finally, herbicide-treated mats create a massive amount of dead matter, which causes a sharp decline in dissolved oxygen levels. These issues notwithstanding, up to 150,000 acres of water hyacinth are treated annually with herbicides in Louisiana.

Biocontrol: Nutria, Manatees, Weevils, and Moths

The third main method is biocontrol. While hippos fortunately failed to be employed for hyacinth control in the United States, another mammal was: a large South American semiaquatic rodent known as the nutria. Smaller than a beaver, larger than a muskrat, and bearing a long hairless tail like a rat's, *Myocastor coypus*—roughly Latin for "beaver rat"—was first brought to the States in 1899 for fur farming in California. Over the ensuing decades, state and federal agents, as well as entrepreneurs, imported and translocated nutria throughout the United States for fur farming and trapping, and the munching critters have been reported in at least 40 states and three Canadian provinces since their introduction. In Louisiana, Edward Avery McIlhenny, noted conservationist and son of the inventor of the Tabasco brand pepper sauce, promoted nutria when he brought 20 of the rodents to Avery Island in 1938 at the encouragement of the Louisiana Department of Wildlife and Fisheries. Because of their prodigious reproductive capability,

A nutria, originally from South America, in Lafitte's Cove Nature Preserve, Galveston Island, Texas. Photo by Robert Nunnally (CC BY 2.0).

within two years his colony had increased in number to more than 500, and McIlhenny began selling them to fur farmers for breeding stock. He eventually liberated this entire stock into the island's marshes to help establish a fur industry for the state. Free-ranging nutria were soon nibbling away at marshes along coastal Louisiana.[27]

McIlhenny's liberated nutria, as well as escapees from other farms, did not cause alarm at the time for two reasons. First, nutria consumed aquatic vegetation and from the 1940s were actively promoted for invasive weed control, specifically water hyacinth. Second, trapping the animals for furs was a boon to the economy. In good times, such as in the 1970s when nutria became the number-one fur crop in the United States, harvests peaked at 1.8 million pelts worth $15.7 million. Wildlife enthusiasts saw nutria as a win-win. The exotic critters would devour the exotic weeds, and their pelts would bring in extra bucks. It was in this light that the East Texas Wildlife Association, a conservation and hunting group, introduced nutria to Caddo Lake after hyacinth became problematic there around 1950.[28]

But the win-win ultimately became a double loss. Nutria will eat the floating hyacinth, but it is not particularly high on their dietary list. They predominantly feed on the base of plant stems and in winter dig for roots and rhizomes. As opportunistic foragers, they exploit a wide variety of native wetland plants, such as pickerelweed, bulltongue arrowhead, cattail, giant cutgrass, and, if available, crops such as rice and sugarcane, at times causing ecological and agricultural damage to the tune of $20 million. They are messy and wasteful eaters, consuming only 10 percent of what they cut down. Their grazing and digging eradicate the marsh's vegetation, exposing large patches of bare mud called "eat outs," sometimes hundreds of acres in size. Naturalists at Caddo Lake noted that a few years after their introduction the nutria were everywhere, but the hyacinth was not reduced. And the nutria fur market turned out to be volatile, with booms and busts over the years. The fur was never popular in the United States—even attempts to market it as Hudson Bay seal failed—and the international market fell sharply, first in the 1950s and again in the 1980s, responding to market saturation, the whims of fashion, and the ethics of wearing fur. When markets crashed, trapping ceased, and nutria populations exploded.[29]

Marsh ecologists now consider nutria, whose densities have been documented at 6,000 per square mile, to be major contributors to the demise of coastal wetlands. Even though nutria meat is higher in protein and carbohydrates and lower in fat and cholesterol than that of most game and domestic animals, restaurants have difficulty promoting it.

Another munching mammal, the American or West Indian manatee (*Trichechus manatus*), was considered a candidate for hyacinth control. The manatee, also known as the sea cow, is a large herbivore (about 10 feet long, 800–1,200 pounds) that thrives in shallow coastal waters, estuaries, creeks, and rivers, tolerating both fresh and salt water. About 6,000 of these slow swimmers inhabit the warm waters of the southeastern United States and range southward (as two subspecies) along the coasts and islands of the Caribbean Sea and the Atlantic Ocean as far as northeastern Brazil. There the manatee is replaced by a smaller species, the freshwater specialist and water hyacinth–loving Amazon manatee.

Watershed managers in Guyana have used manatees for decades to control weeds in irrigation and drainage systems. Although there have been occasional attempts to use the manatee for hyacinth control in the United States in Florida, the manatee is not sufficiently abundant. It is, in fact, an endangered species, and this legal status makes experimentation difficult. In Texas, the West Indian manatee is now exceedingly rare, though a century ago it was not uncommon in the Laguna Madre. The Texas Marine Mammal Stranding Network has recovered fewer than 10 manatees since 1980, although two animals turned up in Corpus Christi and Galveston in 2012.[30]

The most effective and tractable control agents these days are tiny insects that ideally feed on only one thing, the target plant species. In the 1970s

Insect biocontrol agents used to keep water hyacinth in check: (left) a species of water hyacinth weevil (*Neochetina bruchi*) and (right) the water hyacinth moth. Photos, respectively, by CSIRO and Willey Durden, USDA Agriculture Research Service (CC BY 3.0).

the USDA released three such agents for hyacinth control. Two of these, water hyacinth weevils native to Argentina, were first introduced to Florida. *Neochetina eichhorniae* and *N. bruchi* have since become widely established along the Gulf Coast. Weevil larvae tunnel through the leaves to the hyacinth's crown, destroying tissue as they go. Adult weevils then feed on the leaves themselves, which become scarred, curled, and desiccated. This results in an overall loss of growth and vigor.

Another biocontrol agent, also an Argentine insect, was released in Florida in the late 1970s. The water hyacinth moth (*Niphograpta albigutallis*) is now established in Florida, Texas, and Puerto Rico. Its destructive stage involves its larvae, which tunnel through leaf petioles and the plant crown, destroying the growing tip. Damaged leaves and shoots lose buoyancy and sink. The moth larvae proliferate quickly, do damage, and then often disappear, making it difficult to assess the insect as a successful biocontrol. Interestingly, the moth has appeared in Mexico (1993), Puerto Rico (1995), and Cuba (1996) despite no record of any release. Researchers assume the Mexican populations were wind-drifted from Texas, and moths in Puerto Rico and Cuba arrived by a similar means from Florida.

These insects have kept water hyacinth in check, reducing coverage by as much as one-third in Gulf Coast states. The USDA continues to look at new candidates. Between 2010 and 2013, researchers released a planthopper insect (*Megamelus scutellaris*) in Florida, and initial results appear promising. Insects will not eradicate the water weed but will decrease the hyacinth's propensity to reexpand to prior levels of infestation. Insect biocontrol of hyacinth has yielded strong economic incentives. Conservative estimates suggest $40 million annually can be saved by not using herbicides.[31]

Possible Bright Side Benefits

Scientists have explored using water hyacinth as a dietary supplement, notably in Third World nations. Hyacinth fiber has been used in making furniture, handbags, rope, and paper in both Africa and Asia. Recently, researchers have paid attention to the plant as a source of compost, soil fertilizer, or bioenergy. The main problem is that it consists of 90 to 95 percent water. Harvesting green plants, drying, then handling them takes time and space and limits commercial value.

However, water hogging is good for treating wastewater. Given the plant's ability to absorb substances from the water (such as nitrogen, phosphorus, heavy metals, fecal coliform bacteria) and hold them at concentrations thousands of times higher than in the water column, hyacinth become large sponges or scrubbers. In the 1970s many researchers, including the National Aeronautics and Space Administration (NASA) personnel, planted hyacinth in wastewater lagoons. Several cities around the country, including San Diego, Orlando, and Austin, used the plants with varying success.[32] In 2003, when hyacinths invaded the main drinking-water supply of Bogota, Colombia, which had become increasingly polluted, officials found that in just three years the water quality improved to standards originally set for 2020.[33]

In an ironic twist, which is not uncommon in stories about invasive species, Florida's officials have begun *adding* water hyacinth to the waters where for 50 years the state had orchestrated a campaign to get rid of the pest. The waters of Kings Bay in the Crystal River National Wildlife Refuge, increasingly a haven for manatees, have become degraded in recent decades for a variety of reasons, including herbicide use to control hyacinth and hydrilla, reduced freshwater inflow, and increased salinity. These conditions turn out to be perfect for the explosive growth of a slimy green algae (*Lyngbya wollei*), which is unsightly and unbecoming to tourists, and unpalatable and possibly toxic to manatees, which would otherwise be quite happy to feed on hyacinth.[34]

Keeping some water hyacinth around in Kings Bay (albeit in floating cages) reduces algae growth. "If you ignore the label 'exotic' and focus on function," says University of Georgia ecologist Jason Evans, who is advising on the Kings Bay project, water hyacinth "is almost exactly what you would think this ecosystem needs." He continues, "Since these invasive plants are here and we can't get rid of them, I think it's counterproductive to be killing them and not taking advantage of their functions. These are important tools. We should be using them."[35] Robert Knight, a wetland restoration ecologist and director of the Howard T. Odum Florida Springs Institute, reports that the hyacinth's ability to shade-out the green slime and provide a home for snails and other algae grazers is a promising tool for shoreline restoration throughout Kings Bay.[36]

Hydrilla: A Different Approach to Aquatic Invasion

Water hyacinth's comrade in arms is less obvious: it lurks just beneath the water surface. Unlike its floating companion, hydrilla (*Hydrilla verticillata*) pushes its roots into the muck of rivers and lakes. Botanists call it a *submersed* aquatic perennial. And on account of this stabilizing rootedness, plus an adaptability and aggressive growth, many weed experts, echoing the epithets for water hyacinth, regard it as "the most problematic aquatic plant in the United States."[37]

It is important to note that invasive aquatic plants thrive in Texas primarily in unnatural environments. Nearly all of Texas' lakes are reservoirs, that is, artificial impoundments constructed by damming rivers. Streams that once flowed and occasionally spilled over their banks are now still, and their shallow, warm, sunlit bottoms lie beneath deeper, cooler, and darker waters. Dams provide benefits, such as flood control, drinking water, electricity, irrigation, and recreation, but their still waters are a perfect habitat for floating plants. By stopping floods, dams also prevent the periodic scouring that would clear out weeds and other obstructions.

Hydrilla prefers sluggish waters, too, but it especially takes advantage

A typical mass of hydrilla, mostly submerged but with some stems of the plant rising above the water's surface. Photo by David J. Moorhead, University of Georgia, Bugwood.org.

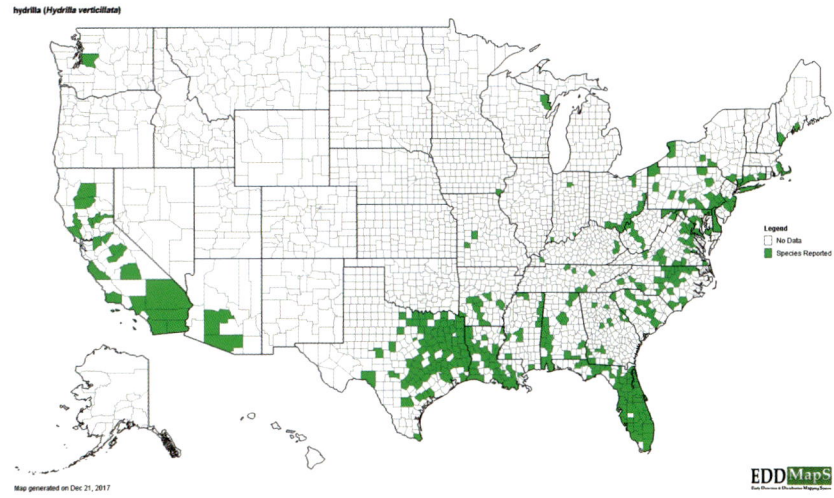

US distribution of hydrilla (*Hydrilla verticillata*). EDDMapS, University of Georgia—Center for Invasive Species and Ecosystem Health, 2017, http://www.eddmaps.org/.

of the depths created by dams. While a reservoir's deep, darker waters are inimical to many native plants, hydrilla thrives because it has a very low light requirement, which allows it to colonize areas inaccessible to other aquatic plants and to enjoy a longer growing season. Needing only 1 percent or less of full sunlight, hydrilla routinely grows on lakebeds 10 feet below the surface and reaches down 50 feet in clear spring water.[38]

Coupling low-light tolerance with 25-foot-long stems and strong growth results in a prolific performer—a monster plant. Like almost all invasive plants, hydrilla grows aggressively, as much as one inch per day on a single stem. This may sound unimpressive, but as the plant heads upward into the sunlit zone, its stems begin to branch profusely. Taking these many branches into account, hydrilla's growth rate approaches 16 feet per day, producing a matlike canopy suspended just below the surface. Even when the plant grows in deeper waters, fully 70 percent of it ends up within two feet or so of the surface, ready to snag and enfold the hapless swimmer or boat propeller. While water hyacinth is a showy floating garden, hydrilla consists of bits and pieces sticking upward here and there.[39]

Like that of water hyacinth, hydrilla's success as an invader is due to its reproductive prowess and ability to spread by fragmentation. In fact, a single leaf whorl is capable of producing a whole new plant, which explains why boat trailers, bait buckets, draglines, and used aquarium water disperse it so easily. But hydrilla has several other reproductive tricks up its sleeve.

One of these is the formation of small dormant buds at leaf bases. Called turions, these eventually mature, break loose, drift, settle, and then germinate. Tubers are hydrilla's next strategy. These small, whitish, pealike swellings appear at the ends of underground stems during fall and winter. One square yard of hydrilla can produce 5,000 tubers, and tuber densities in lakebeds can run into the millions. Since they often grow a foot deep in the sediment, the tubers remain after the parent plant is harvested or dies and are viable for four years or more. When exposed to the air, the sediment blanket protects them from freezes, droughts, and even herbicides. Tubers survive ingestion and regurgitation by waterfowl. In short, one can remove all the visible signs of a hydrilla, but the plant will return.[40]

Hydrilla's final reproductive strategy, though probably not common, is producing seeds. The plant does bear small, inconspicuous flowers. All populations in the southeastern United States, including Texas, are one-sexed (female) and therefore incapable of seed production, but populations in Atlantic Coast states are two-sexed and produce tiny brown seeds, which generally sink into the mud. Seed production adds another layer of protection for the plant's survival and another headache for the people who want to get rid of it.

Threats and Benefits

Hydrilla's main threat is the same as water hyacinth's. Its dense underwater canopy blocks sunlight, displaces native plants, and decreases dissolved oxygen levels, which can result in fish kills. Areas of open water disappear, which tends to reduce the weight and size of sport fish and obstructs feeding areas for some waterbirds. Like the hyacinth, hydrilla mats also disrupt runoff and drainage, increase pooling, and impede the flow of channeled water. Dense infestations of both water plants along the lower Rio Grande, for instance, have slowed flow to irrigators, forcing managers at the Falcon International Reservoir, south of Laredo, to release up to 30 percent more water to push the flow, putting stress on fisheries, farmers, and cities that depend on the lake's waters.[41]

When hydrilla's overall coverage of a water body approaches 50 percent, fishing, swimming, waterskiing, and boating are also all diminished. The plant is particularly harmful to boats, causing the propellers to seize up, and outright dangerous to swimmers by entangling them. This happens to people swimming across the Rio Grande. In the winter of 2014–15, one person per week on average became ensnared and drowned in the weed when attempting to cross the river and its irrigation canals. And in a terrible

Hydrilla stems form dense mats, which may impede boaters and snag swimmers. Photo by Murray, US Fish and Wildlife Service (Pixnio, CCO).

irony, in 1970, Lyle Weldon, a USDA scientist, who helped confirm the first hydrilla infestation in North America, drowned after becoming ensnared in the plant while scuba diving. Weldon was looking for sites near Orlando, Florida, in which he could do his research.[42]

Anglers and waterfowl enthusiasts like this aquatic weed and have occasionally helped spread the plant to *improve* fish or bird habitat. Waterfowl eat this plant, and certain species of predatory fish, such as largemouth bass, use the cover that hydrilla provides to stalk fish prey. Bass anglers in Texas, for example, like hydrilla as long as it is kept in check. Todd Driscoll, fisheries biologist and avid fisherman, points out that anglers in East Texas cast near aquatic vegetation because "it's good cover for adult bass and forage fish and also furnishes an edge that helps anglers locate fish."[43] He deems 20 to 35 percent coverage as ideal. Much beyond this amount and the plant growth works to the angler's disadvantage: it is too hard to get at the fish, and the bass, while abundant, are skinny because the forage fish they rely on can too easily hide. Ideally, coverage consists of native plants, but as these struggle to survive in reservoirs with widely fluctuating water levels and/or deep waters, hydrilla fills the bill.

Introduction and Spread of the "Indian Star-Vine"

Hydrilla originated in the warm regions of Asia and Australia, but it is now everywhere, growing as far north as Lithuania and Siberia and as far south as New Zealand. Hydrilla first turned up in the United States as an aquarium plant. Mistaking it for another aquatic plant, a tropical fish and plant farmer in St. Louis, Missouri, imported hydrilla plants from Sri Lanka in 1951 or 1952. He mailed bundles of them to a plant grower near Tampa Bay, Florida. Unimpressed, the grower threw them into a nearby canal, where the plant grew so well that he changed his mind and began marketing the plant as "Indian star-vine."[44]

Until the early 1980s, it was common for aquarium enthusiasts to raise exotic plants in canals and ditches in Florida. Such water bodies were easy to access year-round and dispensed with the need of a nursery. Outlets in Miami cultivated and sold the Indian star-vine, and by 1960 local waterways had become so choked that the USDA got involved and eventually identified the plant as *Hydrilla verticillata*.

DNA evidence shows that the US introduction, the all-female type, hails from South India. This type spread aggressively throughout Florida and the US Southeast in the 1960s–70s, expanding west into Texas, Arizona, and California. In 1976 the two-sexed type, capable of producing viable seeds, turned up in Delaware and promptly spread into the Potomac River. This fertile type from Korea has continued to spread along the coast, from North Carolina to Maine, and has been found in Washington and Idaho. Hydrilla spreads mainly by fragments attached to recreational boats and trailers, but weed scientists suspect that in the case of California and Washington, tubers may have hitched a ride on mail-order water lily rhizomes.[45]

The worst hydrilla infestations are in Texas and the Southeast. In Texas, where the plant first turned up in 1975, it grows in 110 water bodies, three times more than the water hyacinth.[46] Well distributed in the east—Lake Conroe, Lake O' the Pines, Toledo Bend, Lake Sam Rayburn, to name a few impoundments—hydrilla occurs as far west as Lake Amistad near Del Rio and Lake Nasworthy in San Angelo. Populations of the plant stretch from reservoirs in North Texas near the Red River south to the resacas and oxbow lakes along the Rio Grande.

Methods of Control: Water Drawdowns and Indian Flies

Usual controls involve physical removal and use of herbicides. Physical removal is limited by fragments, such as tubers and turions, that regrow and guarantee swift recolonization. Several herbicides have been tried, but hydrilla is so adaptable that chemically resistant biotypes quickly develop. The plant's resistance to fluridone, a systemic herbicide commonly used on aquatic plants, is already well documented in Florida.[47]

In desperation, officials have begun to implement "winter drawdowns." They lower water levels during winter months, which reduces the space available for the plant in the water column. This slows growth and exposes the plants, as well as uncovers tubers and turions to desiccation and killing frosts. Drawdowns help check hydrilla but do not eradicate it. In fact, a 12-month drawdown in a North Texas lake had little effect on tubers, which apparently never shriveled in the exposed muck.[48]

Biological control of hydrilla is where the excitement and controversy exist. Starting in the late 1980s, two weevils and two flies have been introduced to control the plant. The most successful is the Indian hydrilla leaf-mining fly (*Hydrellia pakistanae*), which has been released in more than 50 sites across Texas, the US Southeast, and California. Native to Pakistan and China, the leaf miner is now established in many sites, between the city of Austin and the Rio Grande, for instance. Tiny fly larvae dine on hydrilla leaves and substantially reduce the plant's ability to photosynthesize. Whole chunks of the plant lose buoyancy and sink. Under optimal conditions, the fly reduces biomass by 30 to 40 percent.[49]

Further Control: A Fishy Tale

To improve results and tackle parts of the plant that grow deep, control experts reach for a larger animal. They use the grass carp (*Ctenopharyngodon idella*), also called white amur, a freshwater fish native to eastern Asia. The chubby, large-scaled, torpedo-shaped fish, silver to olive in color and oddly a member of the minnow family, eats water plants voraciously. A grass carp may consume three times its body weight every day; it grows to three feet long and weighs as much as 20 pounds, though specimens can be much larger. Cultivated for food in China, grass carp have been used widely as weed eaters and were imported into the United States for that purpose in the 1960s. Since the fast-moving rivers needed for their successful spawning are not good habitats for hydrilla, only a few rivers in North America support reproduction. Nonetheless, since the 1980s, fish specialists have added the precaution of releasing only sterile (triploid) grass carp into hydrilla-infested areas. In Texas only sterile fish are legal, and a permit from TPWD is required before purchasing them from a certified dealer.

Using grass carp for controlling aquatic plants has turned out far harder than expected. One of two outcomes is possible: either they have no impact

The grass carp or white amur. Photo by Eric Engbretson, US Fish and Wildlife Service.

on hydrilla, or they consume every shred of plant life, both nonnative and native. The problem lies in calculating how many fish are needed to keep the hydrilla in check. It is a balance between consuming the hydrilla but not all the aquatic flora. Surprisingly, there is no magic number because a number of variables complicate any stocking model. These include local weather conditions; the age and size of the carp to be released; expected migration patterns and mortality; and the size, shape, and location of the release site. Aquatic scientists have used stocking rates from 5 to 175 fish per acre of hydrilla, with an intermediate number of 25–30 fish per acre, to control an outbreak. Sometimes, 8 fish per acre is sufficient to eradicate all the pest plants, while at other times 74 fish per acre appears to make no impression at all. Adding to this imprecision, there is no easy way to recapture fish when it is clear the numbers are wrong. Grass carp are nonspecific herbivores and do not take fishing lures or bait. Sport fishers have no use for them.[50]

Grass Carp: Lake Conroe and Lake Austin

Two examples in Texas illustrate how much finesse and care are needed when stocking carp. Lake Conroe, a 21-mile-long, 20,000-acre reservoir on the San Jacinto River 40 miles north of Houston, was built in 1973 to supplement the water supply to that city. Two years after the river was dammed, hydrilla turned up and spread across 470 acres. Nothing was done. By 1979 hydrilla coverage topped 4,500 acres (22 percent of the lake surface) and had begun to obstruct boating and swimming. Despite initial objections from TPWD staff, the Texas Agricultural Experiment Station, under orders from the state legislature, released 270,000 fertile grass carp into the lake. The calculated ratio was 30 fish per acre of weed. By October 1983 hydrilla, and *all other* aquatic vegetation in the lake, had vanished. The grass carp had eaten everything and continued to feed on any plant that sprouted in the lake for the next 13 years, alarming both naturalists and anglers. Lake Conroe's fisheries managers, the Texas Bowfishing Association, and the Texas Bass Federation Nation have held fishing tournaments aimed at reducing the numbers of the "lawn mowers with fins."[51] Worse was to come after fertile carp left the lake and reportedly began to spawn in Galveston Bay and in the Trinity River that empties into it.[52]

In 1996, hydrilla reappeared in Lake Conroe, and, avoiding the use of carp, water managers sprayed herbicides to check the weed. They also planted 25 species of native plants in the lake. Initially their efforts worked, but within a decade hydrilla again got out of control and herbicides and replanting did not solve the problem. City personnel reluctantly agreed to more grass carp. Hoping to avoid complete denudation of the lake, they copied what seemed a successful example of using carp in Lake Austin, 150 miles to the west.

Lake Austin, one of the seven Highland Lakes of Central Texas formed by damming the Colorado River, is the principal water supply for the capital city. Although less than one-tenth the size of Lake Conroe, it is (or was) one of the nation's premier largemouth bass fisheries. Hydrilla was first discovered in its waters in 1999, growing, not surprisingly, beside a public boat ramp. In 2002, after heavy rains, the sheer mass of the spreading plant began to back up the lake level and flood homes. During this storm, a tangled 100-acre mat floated downstream and caused $384,000 in damage to the trash racks guarding Tom Miller Dam's hydroelectric unit that passes Lake Austin's water into Lady Bird Lake downtown.[53]

The following year, a stakeholder group drew up a management plan for hydrilla. It involved drawdowns, which the Lower Colorado River Authority (LCRA) had already practiced for several decades, and also included *incrementally* stocking *nonreproducing* grass carp. From February 2003 through November 2004, some 8,125 sterile grass carp swam into Lake Austin. Stocking rates varied from 12 to 28 fish per acre of weed. At first the new fish did not seem to be making a dent. But after another flood, which may have washed out some hydrilla, and another 4,675 carp added, experts considered the weed to be under control. Importantly, managers of Lake Conroe liked the fact that grass carp had not denuded Lake Austin.[54]

Lake Conroe officials decided to duplicate what had been done in Lake Austin. TPWD staff and San Jacinto River Association personnel, together with property owners, anglers, and others, devised their own management plan that included incrementally stocking sterile carp. From March 2006 through November 2007 seven releases of about 100,000 triploid carp (from 9 to 42 fish per acre of weed) took place.

The situation worked out well. Late in 2014, officials reported hydrilla had dropped to merely one-tenth acre and that 8,000 grass carp remained in the lake. Native plants had rebounded to cover almost 1,200 acres. The slow, steady approach to solving the hydrilla problem appears to have paid off.[55]

Biocontrol as Balancing Act

Getting the right fish-weed balance is a delicate, precarious, and perhaps impossible task. In Austin, drought conditions throughout Central Texas in 2011 resulted in warmer and stiller waters in the city's lake and sparked a new explosive growth of hydrilla. To combat the weed, fish experts released 17,000 sterile carp to bring the stocking rate to 50 fish per acre. The hydrilla continued to spread, however, covering almost 40 percent of the water body. So in May 2013, officials tipped in another 9,000 carp. That did the trick. Barely four months later, not a single hydrilla plant floated in Lake Austin. "It's not gone, but it is controlled at a level better than we've ever done before," declared Mary Gilroy, environmental program coordinator for the City of Austin's watershed department.[56] The 48,000 grass carp stocked in Lake Austin during the previous 10 years, to the tune of more than $200,000, had turned the tables on hydrilla.[57]

Soon after, however, reports began to filter in that the fish had been too successful. "That lake has undergone an amazing transformation," says veteran Central Texas fish guide Mike Hastings. "They've stocked so many Grass Carp that there's just no vegetation left in that lake. They've eaten everything, even the bulrushes."[58] Bulrushes and other plants bordering the lake are habitat for waterbirds and nurseries for fish, and they buffer the wakes of powerboats. All plants had disappeared, and as grass carp live for 10 years or more, the prospects for the plants coming back soon are remote. Lake Austin is now a degraded water body, not only in the loss of edge habitat but also for the amount of bank erosion that is happening along its exposed shores. The bass fishery has collapsed.

Grass carp have other downsides. Apart from posing a reproductive threat (when not made sterile) and being voracious consumers of water plants (native and nonnative alike), the fish digest only about half what they consume. They expel the rest. Fish waste increases nutrient loads, lowers oxygen levels, and may increase algal blooms, which, in turn, lower water clarity and the oxygen content in a feedback loop. The fish also harbor parasites. One of them, the Asian tapeworm (*Bothriocephalus acheilognathi*), which lives in the gut of a variety of freshwater fishes but particularly those of the carp family (Cyprinidae), has spread worldwide, decreasing the size of fish it infects and causing economic losses to fish farms.

These examples from Lake Conroe and Lake Austin clarify two things. First, finding precise and effective stocking rates for biocontrol fish is as much an art as it is a science. Experts recognize that it takes time and patience to arrive at an estimate because the number and types of variables are neither completely known nor well understood. Second, hydrilla will never be eradicated in Texas or the United States. Rather, its presence underscores the need to stand back and find ways of planning that use carp-type fish with other control agents. Such thinking involves cooperation among stakeholders and enough constructive discussion to get the timing and releases right. "Carp [keep] hydrilla mowed to the bottom of the lake, much like cows chewing on a pasture. But carp don't kill the plant entirely," Gilroy says, referring to the tubers in the sediment that are bound to resprout.[59]

Referring to carp as livestock makes an interesting analogy. We introduce a nonnative grazer to keep in check a nonnative pasture. This always seems a risk, if not in theory then certainly in practice. If the grazers do their job but overgraze the pasture, they will turn to whatever forage they can find. While using sterile grass carp may be a cost-effective method for hydrilla control, the fish is recommended only for closed water bodies such as ponds, lakes, and canals where the complete denudation of vegetation may be tolerable.

Turning back to the Hippo Bill of 1910, one wonders what the downsides would have been if we had imported the 1.5-ton behemoths to consume vegetation for which they had no actual appetite. Would they have devoured water hyacinth out of necessity, or would they have instead competed with cattle in bankside pastures? Would we have developed a taste for hippo brisket? The story of nutria suggests mainly negative outcomes for these questions. In the case of grass carp, the final irony is that the fish, along with other Asian carp, are perfectly edible and have been a staple in Asia for generations. Unlike the bottom-feeding common or European carp (*Cyprinus carpio*), whose meat is bony and often muddy tasting, grass carp are like other mild white fish, though they are bony. The only things keeping us from eating them are the difficulty of catching them and the name "carp," which people associate with the bottom-feeding, mud-rolling species that was tried as a food source a century ago and then largely rejected as a trash fish. People seem to reject nutria as an entrée for equally capricious reasons.[60]

Water hyacinth, like sparrows and starlings, came to our shores before the passage of the Lacey Act. We invited it into our garden ponds as a curiosity and for sheer aesthetic delight without the slightest foreboding. No one could have supposed that an aquatic herb would bring river navigation and commerce to a halt and wreak havoc on ecosystems within a decade of its arrival. Hydrilla shows that a half-century later a much less distinguished freshwater plant, renamed to promote its desirability, repeated the story. In both cases we have inadvertently assisted the spread of these invasive species by providing them with both natural and humanly crafted environments to exploit. Unlike the case of the two birds, we attempt to keep these aquatic plant populations in check by importing other nonnative creatures as control agents. Biocontrol agents have been partly successful, but their appearance in places far removed from release sites (in the case of water hyacinth moths) or their ability to devour native nontarget plants (in the case of grass carp) are serious problems. Will we end up having to control the controller as we do with nutria? Finally, these two plants remind us that even major invasive species may provide ecological benefits, such as filtration (water hyacinth) and fish habitat (hydrilla). This encourages us to contextualize our strategies as we seek to manage them.

3 Feral Hogs
Pork Chops or Sherman Tanks?

Feral hogs have been in Texas a long time. Under Spanish rule, settlers introduced domestic swine that soon ran wild. Landowners did not complain about a domesticated animal that subsisted around farms and ranches. By the middle of last century, many hunters regarded feral hogs as game animals and introduced some for sport. However, experts explained that hogs damaged crops, pastures, parks, and suburban spaces and collided with vehicles. Officials supported programs to reduce numbers by shooting, trapping, tracking with hounds, killing from the air, and setting out toxicants. Recent trials involve a common, low-cost chemical dispensed to kill hogs in a humane way. However, hog experts conclude it is impossible to eradicate wild swine in Texas. Strategies to stabilize and perhaps lower numbers are limited by hog biology and by the ways they adjust to humanly transformed environments.

Hog trap, an effective and traditional way of controlling feral hogs. Photo by US Department of Agriculture (CC BY 2.0).

Jim Ellender aims at a group of wild, multicolored pigs grunting and snuf-
fling 50 feet or so from his blind. He has a perfect shot. They are sitting targets, so he squeezes the trigger. Instead of hitting a feral pig, his bullet pings
off a plate that releases the door of a wire mesh corral into which a line of
corn has lured them. Habituated, the swine have grubbed the bait and are
now trapped. His bullet slams the door shut, leaving nine panicked animals
dashing around the circle of so-called hog wire.

Jim, manager of a ranch in the Texas Hill Country, has two options. He
can shoot the pigs, field-dress one or two for family and friends, and leave
the carcasses for scavengers. Or, as he does, he can back up a trailer to a
cage, herd in the beasts, and haul them off to a holding site. The operator
of the facility will take his live pigs and hand him a check. After a sufficient
number of feral hogs have been delivered, the holding facility will pass them
on to a slaughterhouse, where they will be inspected for diseases, notably
pseudorabies (the infection rate is about 20 percent) and swine brucello-
sis (about 10 percent are infected), butchered, and sold on to wholesalers as
far away as Asia. Unlike venison, which hunters can donate to charities and
food banks, wild pork has to be checked. A company founded in Austin in
2016 sells harvested hogs as pet food. The Texas Animal Health Commission
approves the holding facilities (there are about 125 of them) in which wild
hogs weighing 60 pounds or more may be held until taken for slaughter.
The commission stipulates that feral hogs must be kept at least 200 yards
away from domestic swine.[1]

Jim Ellender is following a policy recommended by state biologists and
agents as a way of coming to grips with a plague of hogs estimated to be as
many as 2.5 million, but nobody knows for sure. The Lone Star State has the
dubious distinction of possessing half of an estimated 6 million feral swine
in North America. Texas AgriLife Extension Service experts say feral hogs
occupy at least 90 percent of Texas, that is, 230 of the state's 254 counties,
and are most abundant in river bottoms, thickets, and croplands in eastern
and central areas. They are also at home in the South Texas brush country
and are rooting in the borderland with Mexico. Free-roaming hogs, which
average 100 pounds or more and grow three feet tall at the shoulder, are
causing at least $50 million in direct damage annually throughout the state.
This is a fraction of an estimated $1.5 billion worth of damage across the
nation. However, losses in Texas are climbing as its feral hog population,
ranked second to that of native white-tailed deer, multiplies and spreads.
This is also happening in other states (36 have documented populations)
and four Canadian provinces.[2]

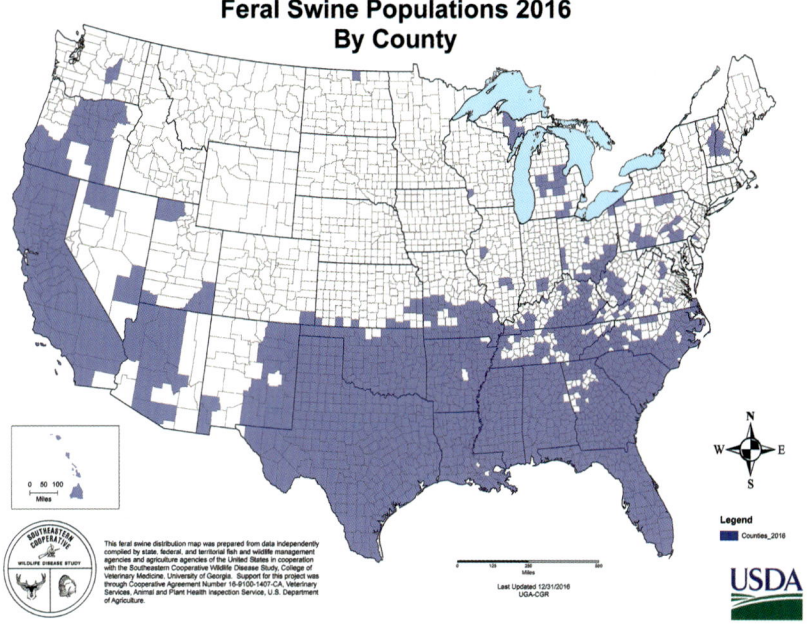

Feral hog distribution in the United States by county, 2016. Photo by US Department of Agriculture, APHIS.

In an article about these marauding pests in the *New York Times*, Erica Goode notes that in Michigan, for example, where numbers of wild boars and hogs on game ranches have increased, individuals escape or have been freed for hunting. The boars will breed with feral hogs. Swine numbers are growing in New England and across the High Plains, such as in the Badlands of North Dakota, where a multiagency task force has been eliminating them. Hog populations are continuing to boom. The USDA's Division of Wildlife Services has recently asked for $20 million to tackle a national hog problem.[3]

The issue with hogs as with other domesticated animals, such as horses, burros, and house pets, is that once they get away from people, the animals prove smart and wily and breed well. Feral hogs prove hard to manage and impossible to eradicate. They also have a reputation for turning aggressive after being wounded, threatened, or cornered. Old-timers say, "If hogs knock you down, likely they'll eat you." Among the animals in this book, feral hogs are special because settlers originally depended on their ancestors, rendering carcasses for meat and body parts, after letting them roam.

There is in theory no distinction between the usefulness of the domestic pig that has been penned and tended for upward of 10,000 years in the Old World and the feral pig that has been rooting in Texas for at least 300 years, except dealing with the feral ones is increasingly difficult and expensive. It is said that it is better to butcher and cook younger ones that weigh less than 200 pounds, as they taste better.

Hispanic Hogs

The hogs in question originated from eight domestic animals Christopher Columbus reportedly purchased on La Gomera, second smallest of the Canary Islands, and put ashore on Hispaniola (today's Haiti and the Dominican Republic) on his second voyage in 1493. Spanish-introduced swine reportedly multiplied and ran wild throughout the Caribbean, becoming seed stock for populating Hispanic America, including Mexico, Panama, and Colombia. A 1506 report noted that colonists had begun to hunt wild pigs that were already rooting up their crops.[4]

Hernando De Soto's expedition (1539–42) carried pigs born in Cuba to the US mainland. Expedition members trailed the herd out of Florida. De Soto bartered some, gave away others, lost more, or had people steal them. By various means, the porkers, originally from the Old World, gained a hoof-hold in the American South. After De Soto died near present Fort Smith, Arkansas, in 1542, Spain's Estremadura-born Luis de Moscoso Alvarado assumed command. He headed north and west with his herd of swine, reportedly totaling 700 animals, and may have closed on the Brazos River in Texas before turning back to the Mississippi River. Once there, survivors reduced to about one-half the 600 or so men who had disembarked from nine ships four years earlier, fashioned crude boats, and floated downriver. After hugging the shorelines of today's Louisiana and Texas, the crew disembarked in Mexico and eventually reached Mexico City, having left what hogs remained to fend for themselves.

Later, explorer Robert de la Salle's tragic and misguided landing on Matagorda Bay (while looking for the mouth of the Mississippi River) in February 1685 added more pigs to the Gulf Coast landscape. Whether these animals proved significant over the longer term we do not know. Eighty years or so before La Salle, additional domestic swine had already swum and stumbled ashore, about 1,400 miles to the northeast. They were livestock for John Smith's English settlement in Jamestown, Virginia. Pigs began to trot inland from the Atlantic Seaboard as well as the Gulf Coast.[5]

One thing is clear: there have been ample opportunities for domestic swine to have established themselves and spread throughout the US South, including Texas. Hog specialist Richard Taylor notes that by the early 1800s hogs were reportedly numerous around Nacogdoches, Texas. An 1834 census listed 60,000 animals, almost seven times more than their human keepers. About that time, a statistical report by Colonel Juan Almonte to Mexican authorities noted that 21,000 people (excluding Native Americans) lived in the three Mexican Departments of Texas. The residents owned 110,000 hogs, more than sheep and cattle combined.[6]

From Livestock to Game Animal

Under the dozen or so empresario land grants in the 1820s and 1830s pioneered by Stephen F. Austin, settlers customarily herded and marked the swine of mixed ancestry before letting them roam. Residents regarded hogs as privately owned and used dogs to round up some periodically to sell or slaughter. Only after landowners began to fence properties and laws bolstered enclosure did Texans treat feral hogs as wild game. Wealthier newcomers also introduced European wild boars, prized for hunting. Inevitably, some boars interbred with local hogs and added to the gene pool for recreational hunting.

Brian McCombie notes that the image of the feral hog as a game animal became common. In fact, from the 1970s through 1990s, enthusiasts transported and released more and more of them, he notes. In addition, land-

Feral hog damage to a yard in suburban Tyler, Texas. Photo by Billy Higginbotham, Texas A&M AgriLife Extension Service.

Hog wallows erode and pollute the edges of ponds and stock tanks. Photo by Robin Doughty.

owners started to put out feed for deer, turkey, and quail. Access to such nutritious handouts contributes to a sow's fecundity, he explains. Biologists and wildlife officials complained that hogs were rooting up pastures, damaging native animals and plants, and polluting ponds and marshlands. State agencies urged landowners to halt the illegal practice of transplanting hogs and to treat them as invasive pests.[7]

Currently, both the Texas Parks and Wildlife Code and Texas Agriculture Code list feral hogs as exotic livestock (neither game nor nongame) that belong to the landowner whose property they inhabit. If hogs are causing damage, the owner can hunt them anytime day or night throughout the year and kill as many as is desirable—without a state hunting permit. A hunting permit is required only if the owner intends to snare, trap, or kill feral hogs for meat or trophies.

Such open-ended opportunities attract sport hunters and ranchers to generate income by charging an entry fee and extras for every hog killed. However, to keep hog numbers in check, notes professor and wildlife specialist Billy Higginbotham, today's shooters have to kill 60–70 percent of feral hogs every year. Recreational hunters take perhaps 25 percent, leaving the remainder to double every five years or so. We have more than we can deal with, declares Higginbotham, who reminds land managers that if they do not have hogs now, they soon will.[8]

Congregating around available water shows how adaptable, fecund, and protective feral hogs can be. Photo by University of California's Division of Agriculture and Natural Resources, Sedgwick Natural Reserve.

The biological storm that experts describe depends on the hog's ability to breed frequently and have large litters, especially after feeding around deer feeders. A sow can drop four or five piglets when six months old, though she is usually older. She is a good mother and, with a female helper or two, can protect and raise her piglets. She can also bear two litters every 12 to 15 months. Such fecundity, gregariousness, invulnerability to predators (after about three months old), and access to a wide range of foods explain the current population boom.[9]

Helicopters and Other Control Techniques

With a landowner's permission, Texas law allows feral pigs to be shot from the air, including from helicopters. This so-called pork chopper bill (H.R. 716) introduced by State Representative Sid Miller of Stephenville, gained broad support and received then-governor Rick Perry's signature in May 2011. The law permits licensed hunters to hire helicopters and contract with landowners to shoot feral hogs and coyotes on their properties. Costing between $300 and $600 per hour, gunning from the air may have a special appeal in South Texas, where a lot of scrub and brush is home for a lot of wild hogs but few people.

However, Austin's outdoor writer and hunter Mike Leggett chastised the helicopter measure as merely shooting, not hunting. Leggett doubted whether hog hunters would be prepared to pay four or five times more for blasting away from a helicopter. Gunning from a helicopter may dent numbers while pandering to gory videos about animals dying, he declared.[10]

Neither Leggett nor TPWD personnel have a ready solution. The invasive species is spreading into metropolitan areas. Most efforts concentrate on controlling, snaring, trapping, and shooting and in rousting animals from places that people want to occupy. In an urban wildlife conference held in Austin, Rob Denkhaus, manager for the Fort Worth Nature Center and Refuge that holds one-third of the parkland within that Texas city, talked about his battle with hogs. It had been going on for years, after feral pigs of unknown origin, he noted, turned up on the 3,600-acre space in Fort Worth. Denkhaus and colleagues set about removing 227 wild pigs from the refuge, mostly by corral-trapping them. Compiling body counts, claims Denkhaus, is futile because "eradication is unrealistic." The best one can do, he insists, is to reduce their presence to acceptable levels.

The problem of feral pigs in Fort Worth surfaced in the late 1990s when officials did not share the manager's growing concerns. Using greenbelts and drainage corridors, animals slip into communities. The spread of hogs in the Dallas–Fort Worth Metroplex is happening in other cities as far apart as San Jose, California, and Detroit, Michigan, and is becoming an urban problem globally. Hogs mangle sprinklers, dent and damage vehicles, scatter garbage, overturn tombstones, and even enter empty buildings. An estimated 3,000 boars are inside Berlin, Germany, and some clatter into open houses.

In October 2012, when an eastward extension of toll road SH 130 with a speed limit of 85 miles per hour opened around the city of Austin, several vehicles collided with hogs (fortunately, none caused life-threatening injuries to the drivers). Police officers observed herds scurrying across the highway. The toll road bisects farm- and ranchland in Caldwell County, where officials placed a bounty of $5 per hog. By early 2013, more than 1,000 animals had been destroyed in Hays and Caldwell Counties. Bastrop County, east of the capital city, offered a $5 bounty on hogs in 2013. For payment, hunters submit hog tails to designated feed stores.[11]

Which controls are working? Biologists with TPWD and other agencies rely on trapping and hunting. Trapping is the most common technique for which plans and instructions are available on websites through the state's Extension Service. Some hunters spotlight or use night-vision optics to shoot animals after baiting them with corn. Others capture and radio-tag a feral hog, often a female, turn it into a so-called Judas pig, and track the signals to other members of the sow's group. Everyone agrees, however, that the Judas technique will never eradicate the pest.[12]

A long-standing way of hunting is to set dogs on feral hogs. John Morth-land describes how bay dogs and control dogs work in South Texas. Bay dogs, often foxhounds, catch a hog's scent, and then Rhodesian ridgebacks, Catahoula curs, or black mouth curs (the working dogs) start to follow, barking loudly. "We like a dog with a medium-cold nose, silent on trail, a loud mouth that is easily located, a dog that is fast enough to outrun any woods rooter, strong and quick enough to bay up a hog at his head until the catch team arrives," declares a promotional website for the foundation Black Mouth Cur Breeders Organization in Texas. Baying black mouths announce their hog, chase and corner it, darting away from razor-sharp tusks. Catch-ing up, the hunters unleash the catch dogs, maybe pit bulls (that may be outfitted with Kevlar vests, notes Morthland), to tackle the hog physically until humans get close enough to kill it.[13]

Another answer about what is working comes from half the world and another hemisphere away: Australia—home for many invasive animal spe-cies, mostly introduced from Europe. Feral hogs are a menace in that South-ern Hemisphere continent. Estimates suggest there are probably more feral hogs (23 million) than humans in Australia, and they are doing at least 100 million Australian dollars in damage every year. In desperation, the Aussies turned to three toxins, each with its own problems, to poison wild pigs—so-dium fluoroacetate, yellow phosphorus, and warfarin.

Sodium fluoroacetate (known as compound 1080), for which there is no antidote, has humane issues that include pain and secondary poisoning of nontarget animals, including livestock and pets. The chemical toxicant also persists in the environment. This unintended effect also plagues other toxicants, including yellow (or white) phosphorus. Yellow phosphorus is a widely used poison against rodents. Warfarin, an anticoagulant often used to kill rats, is effective, but individual animals vary in the quantity they consume before it kills them; an individual may take several days to die. State and government agencies in Australia regard yellow phosphorus and warfarin as inhumane and unacceptable. However, in 2017, Texas Agricul-ture Commissioner Sid Miller promoted a warfarin-type poison for wild pig control. In response to the threat of lawsuits, the company licensed to apply this EPA-approved poison withdrew its registration request for appli-cation in the state. Opposition to warfarin centered on the need to assess longer-term environmental and the humane issues of using it (as did rat poison approved in 1952, and a blood thinner in humans in 1954) on feral hogs.[14]

Sodium Nitrite: The Silver Bullet?

Australia's Invasive Animal Cooperative Research Centre, the nation's largest integrated invasive animal research program, has developed a new toxin: "Hog Gone," which is made of sodium nitrite, a common food additive. Sodium nitrite is a relatively inexpensive, widely available methemoglobin-forming compound, which prevents red blood cells from releasing oxygen to the brain and vital organs. Ingesting sodium nitrite induces lethargy and a loss of consciousness in hogs. After eating it, they go to sleep and never wake up, so wild hog experts regard the toxicant as humane and effective. The chemical is highly toxic to pigs because they lack the natural enzyme that reverses its effects: it takes about one and a half hours for most hogs to die after ingestion. Experts compare the compound's action to carbon monoxide poisoning—lowering suffering by putting the individual to sleep. The chemical also degrades quickly and has low bioaccumulation and secondary poisoning. An Australian government Veterinary Services Division Study concludes that "the behavioral evidence suggests that death from nitrite intoxication is an acceptable method of humane kill-

The Hog-Hopper, a swine-specific bait delivery system being tested in Australia in collaboration with US Department of Agriculture APHIS researchers. Photo by Tyler Campbell, USDA APHIS (CC BY 2.0).

ing for large scale feral animal culling." When a delivery method is approved in the United States (Australia does not have additional mid-sized mammals to worry about), probably a bait-holding hopper with a heavy metal door that a hog can lever up with its snout, sodium nitrite will significantly add to the possibilities for control.[15]

Sodium nitrite is more lethal and more humane than compound 1080. The toxicant reduces nontarget impacts because wild animals, other than hogs, are less sensitive to sodium nitrite, for which there is also an effective antidote, methylene blue (used to fight malaria and cancer in human beings). Only scavengers that directly feed on the stomach contents of poisoned pigs may be at risk. However, there is the possibility of Hog Gone baits spilling into rivers, ponds, and lakes where sodium nitrite is lethal to fish.[16]

US field trials using sodium nitrite started in 2010 by placing nontoxic baits in sturdy feed hoppers. The USDA's Division of Wildlife Services conducted hopper experiments in six states, including Texas. Useful data have been generated from trials in upland, wetland, woodland, grass, and brush habitat in 13 counties. These habitats support white-tailed deer, collared peccaries, coyotes, bobcats, and badgers, as well as feral hogs. Researchers found that only raccoons learned how to raise the guillotine-type doors that hogs found easy to open.

The Captive Feral Swine Research and Demonstration Facility at the Kerr Wildlife Management Area has been an important actor for sodium nitrite. Hog hoppers in different sites under a range of conditions enable experts to check various bait ingredients, how many the hogs consume, and how quickly the substance degrades. The toxicant does not withstand high temperatures, for example.

Experts also tested whether consuming pork with sodium nitrite affects hunters and looked at how scavengers react on sodium nitrite–poisoned hogs. Generally, results support the use of this new bait. Experts in Louisiana, where at least 500,000 feral hogs are rooting, have devised an ingenious delivery system. Upon detecting the grunts of feral hogs, a hopper door opens. First, however, the device communicates pictures of the hogs to a cell phone. Once checked, a command opens the hopper door, allowing the hog's access to the toxicant.

The USDA has not approved this device. Some biologists wonder why the state agricultural commissioner championed warfarin without waiting for the results of using sodium nitrite. John Kinsey, research biologist at the Kerr Wildlife Management Area, claims high rates of success with this more humane toxicant; additional ones are planned. Kinsey says he is "pretty confident about the low level risk to non-target species" and that Texas Parks and Wildlife has a good chance of targeting delivery to feral pigs.[17]

As one of the 100 world's worst invasive species, the feral hog is a global menace. Sodium nitrite is likely to be the eventual winner. The USDA is working out the kinks to make hoppers and baits containing this toxicant as species limited as possible. Tyler Campbell, feral swine project leader for USDA Animal and Plant Health Inspection Service (APHIS), notes that results of tests and trials with the new toxicant are encouraging in respect to targeting wild hogs as specifically and as humanely as possible. Can this be the silver bullet?[18]

Topping 28.5 million in 2017, the number of Texans has doubled since 1980. Population growth from within and immigration from neighboring states, such as California, has sustained an increase that ranks Texas the second most populous state in the nation. Most older and newer Texans live in metropolitan counties. Houston, Dallas–Fort Worth, Austin, and San Antonio have blossomed with suburban developments and edge communities. More and more of these residents are encountering feral hogs in county parks, municipal golf courses, river retreats, and even highway medians and verges that have pushed into and transformed rural areas. Residents with vacation and second homes located on scenic overlooks and lake edges also witness firsthand the impacts of what some have termed "mini–Sherman tanks" have on the landscape and object as never before to the economic and ecological damage they inflict. As the number of Texans continues to grow, more residents experience or are made aware of the damage done by the long-ago domesticated animals and support both local and state programs that reduce and contain it.

4 Chinese Tallow
Unrealized Promise

The greatest service which can be rendered any country is to add a useful plant to its culture.
—Thomas Jefferson, "Summary of Public Service"[1]

Although brought to our shores more than two centuries after domestic swine, Chinese tallow is among the earliest imported nonnative trees to become invasive. Like the sparrows and starlings, the tree was carried to our shores for its perceived benefits. In fact, one of America's Founders, Benjamin Franklin, specifically regarded the tallow as a crop tree for producing oil. We encouraged the tree's expansion until long after its introduction. Sparrows, starlings, water hyacinths, and hydrilla created obvious problems soon after they arrived. But Chinese tallow, like feral hogs, managed to stave off criticism despite proliferating. We overlooked the alien tree's dubious role because it looked pleasing; supplied an unexpected resource, honey; and withstood our hope of becoming a source of valuable oil. The cost of that hope has been habitat degradation and the loss of coastal grasslands.

(left) Chinese tallow as a yard tree in fall foliage and (right) close-up of leaves, some fruits noticeable. Photos by Matt Turner.

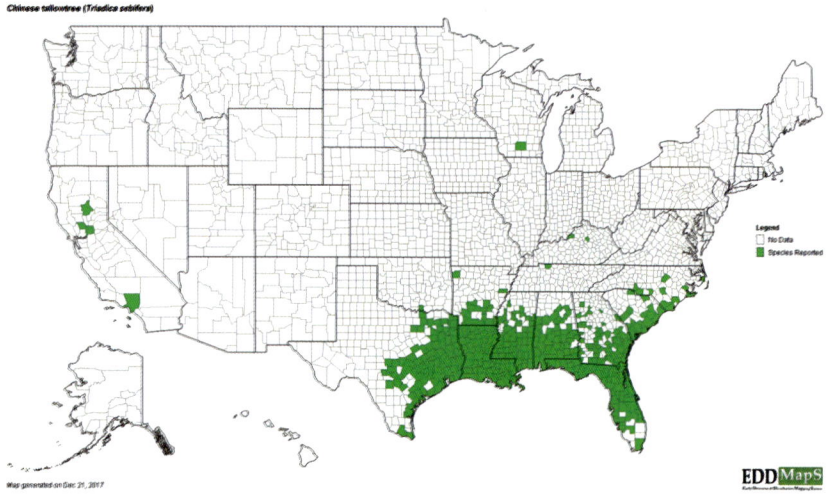

Chinese tallowtree (Triadica sebifera)

Legend
No Data
Species Reported

EDDMapS

Map generated on Dec. 21, 2017

US distribution of Chinese tallow (*Triadica sebifera*). EDDMapS, University of Georgia—Center for Invasive Species and Ecosystem Health, 2017, http://www. eddmaps.org/.

Along the estuaries and levees of southern China there grows an ornamental tree with spectacular fall foliage. Each autumn, the leaves of the *u-kau-shu* or *wu-yau-shoe*, as the tree is called in Cantonese, blaze with orange and red, burnishing to vermillion and mottled maroon. As winter approaches, the leaves fall, exposing thousands of white, pea-sized seeds. These are attractively arranged in little groups of three, or triads (hence the genus name *Triadica*). Bits of the brown husks that once contained them still partly cling to the stems. The overall effect of the white and brown clusters earns the tree the fanciful moniker "popcorn tree." Many of these clusters persist into the winter, making the tree attractive in all seasons.

Eleven thousand miles away on the Texas coast this same tree is now "the single greatest threat to the continued existence of coastal prairie" according to wetland researchers.[2] Dense, impenetrable woodlands, comprising almost exclusively this exotic tree, are marching across the prairies of Southeast Texas. Popcorn tree, or Chinese tallow, as it is known most commonly in the English-speaking world, has moved well away from its hearth in Asia to become the most abundant tree species in an eight-county area around Houston, accounting for almost 25 percent of all trees in that region. South and southwest of Houston—Texas' largest city and fourth largest in the United States—less than 1 percent of the original coastal prairie remains. From an ecological perspective, the arrival and encroachment of the tallow

(left) Chinese tallow in full fruit. (right) As fruits ripen and dry, they split open to reveal triads of waxy seeds, which contain both saturated and unsaturated fats that are potentially valuable commodities. Seeds with split pods still attached give rise to the fanciful name "popcorn tree." Photos by Matt Turner.

tree into that grassland is practically the final nail in the coastal prairie's coffin. Along with suburban growth, rice farming, and oil and gas development, Chinese tallow is yet another intrusive element to radically alter the face of the Gulf Coast lowland and the Magnolia Capital, Houston.

Chinese tallow's popcornlike seeds are key to its presence and now-forgotten fame. These seeds are loaded with waxes, or fats, to which the tree's species name, *sebifera* (bearing wax), refers. They are unusual in containing two distinctly different fats. The white substance surrounding the seed is a saturated fat, or vegetable tallow, similar to cocoa butter. This tallow renders into soap, candles, and other waxy products. In fact, wax from tallow seeds has been used over the centuries in China to make candles for Buddhist ceremonies, which burn with a clear, clean, smokeless flame. This is the reason that tallow trees have been cultivated in several provinces of that country. In southeastern China the tree was so valuable that locals in Hangchow used to pay their taxes with the seeds.[3]

The seed kernels themselves contain an unsaturated fat, which has different applications. The fatty acids of the kernels can be extracted into a drying oil, similar to linseed. This pale yellow kernel oil, called stillingia oil after an older name for the plant's genus, hardens to a tough, solid film. It, too, can be burned like the outer wax for illumination, but it also makes an excellent drying agent in varnishes, paints, lubricants, cosmetics, and even in plastics.

Dispersal and an Auspicious US Introduction

Humankind is always on the lookout for natural resources, and the potential to harvest one, if not both, of the tallow oils encouraged the tree's dissemination around the world. The tree's original distribution likely centered on, or lay just south of, the Yangtze River valley and included waterside habitats throughout southern China, northern Vietnam, and probably Taiwan. But over the last 250 years, the tree has spread mostly with human assistance into Japan, Indonesia, Java, Singapore, India, Bangladesh, and Pakistan. Enthusiasts planted tallows along the coast of the Mediterranean Sea in France, Italy, and Algeria. They set seedlings in Mexico and on Caribbean islands, such as Puerto Rico, Cuba, Martinique, and Jamaica. Tallow trees are thriving in Costa Rica, Panama, and Peru and are even established along the eastern coast of Australia.

In the United States, Chinese tallow enjoyed a particularly auspicious debut, for it was none other than Benjamin Franklin who first introduced the colorful tree to our shores. Franklin lived in London in the 1750s and acted as agent for several American colonies, including Pennsylvania and Massachusetts. While in England, he became acquainted with a John Ellis (1705–76), an Irish naturalist and fiscal agent for the British colony of West Florida. Ellis was an enthusiastic amateur scientist in the best of the Enlightenment tradition and had been inducted into the prestigious Royal Society in 1754, two years before Franklin.[4] One of his passions as a naturalist lay in spotting plants useful in horticulture and developing techniques to ship seeds and specimens long distances. With the good fortune of having the colonial governor of Georgia as his nephew, Ellis sent things like licorice roots, rhubarb, and English acorns to Georgia and, in return, received botanical shipments from the New World in England. He also had agents in Southeast Asia routinely ship him specimens. Franklin, who almost certainly acquired Chinese tallow seeds from Ellis, sent them to his main correspondent in Georgia, Noble Wimberly Jones, a physician and prominent statesman in Savannah. Franklin tells Jones in his letter of October 7, 1772: "I send also a few seeds of the Chinese Tallow Tree, which will I believe grow & thrive with you. 'Tis a most useful Plant."[5]

This kind of casual exchange of plants was a common practice among entrepreneurs and well-heeled landowners on both sides of the Atlantic Ocean. Innovators were interested in novel and potentially valuable products. Jones sowed Franklin's tallow seeds on his Wormsloe plantation, the oldest of Georgia's tidewater estates, likely the following spring. Today the site contains the ruins of the original fortified house. It is the oldest structure still standing in Savannah, and importantly the descendants of Jones's tallow plantings still grow there. As far as is known, Jones's planting of Franklin's seeds is the first introduction of the Chinese tallow tree to the United States.

The tree dispersed from Jones's plantation, but it did not become invasive as we might have predicted; this resulted from subsequent introductions. In these early years the only state other than Georgia in which tallow appeared was neighboring South Carolina. Alexander Garden of Charleston received tallow seeds from Ellis a few months after the ones Franklin had supplied to Jones. Garden reported that his seedlings survived the winter and were thriving; in fact, the ones planted outdoors were looking healthier than those planted indoors. Ten years after this planting, Bavarian-born physician and naturalist Johann David Schöpf noted that Chinese tallow was doing well along Charleston's seacoast where hard freezes from the interior did not penetrate. Thirty years later, in 1824, according to Stephen Elliott, the tree was "completely naturalized" in the coastal zone in both Georgia and South Carolina. Known as the "father of southern botany," Elliott portended more than he realized when he noted that tallow seeds, "though they contain much oil, no use is yet made of them."[6] Half a century after its introduction, Franklin's "most useful plant" was not being put to use at all.[7]

The Real Invasion Begins: Human Promotion
Beyond its core region, Chinese tallow grew sporadically in the South. Planters admired it as an ornament, something unusual and different in a garden or property. In New Orleans, Scottish botanist and collector Thomas Drummond happened upon a specimen in 1832. Despite such sporadic appearances, the real invasion of Chinese tallow came later and from elsewhere. In the early 1900s, the Foreign Plant Introduction Division of the USDA's Bureau of Plant Industry began to promote the cultivation of tallow for a soap industry. Sizable plantings took place near Jacksonville, Florida, and Houston, but commercial efforts failed apparently because of the prohibitive cost of harvesting the seeds.[8]

People would have quickly noticed the lovely tree. Edward Teas, who founded Teas Nursery in Bellaire, Texas (at the time, a suburb just west of Houston), is said to have introduced Chinese tallow in about 1910. Since he moved his nursery to Texas from Missouri in that year, he probably did not actually introduce the tree to the state but likely propagated and promoted it. Teas and his descendants helped landscape Bellaire, Rice University, and the Montrose and River Oaks areas of Houston.[9]

The desire for shade helped spread Chinese tallow. Brownsville residents were growing the tree for shade in the Rio Grande Valley by 1925. The tree's ability to form a closed canopy within a mere 20–25 years doubtless worked in its favor. In the 1940s, Texas Highway Department personnel planted tallow along the highways as shade for passersby and roadside livestock (the tree's toxicity to cattle was discovered only later). Supporters touted it as "an ideal tree for the vast, treeless expanses along the Texas Gulf Coast, as it would provide shade for the livestock, as well as produce a crop of seed of considerable value."[10] A decade later, residents throughout the Gulf Coast states planted the tree for both shade and ornamental purposes.[11]

Despite almost two centuries of failed attempts to make Chinese tallow economically viable, hopes for an oil crop continued. Melvin Handley, a chemist working for Dow Chemical Company in Freeport, Texas, experimented with tallow as a hobby and envisioned the tree as a cash crop for southern agriculturalists. So in 1949, Handley distributed thousands of seedlings to farmers across a 600-mile inland belt of the Gulf Coast, hoping that they could take advantage of the oils (for cosmetics, plastics, and floor wax) or, at the very least, could grind up the seeds for fertilizer or chicken feed. In fact, the tree is commonly planted near chicken coops in Louisiana, where some people call it "chicken tree."[12]

These twentieth-century plantings, and doubtless many others, appear to be what sparked a stampede of tallow expansion. DNA evidence reveals two lineages exist. The first comes from trees planted in the lowlands of Georgia and South Carolina, the site of the tree's original introduction. The second includes trees in the remainder of the US Southeast. In short, Ben Franklin and his cohorts are not to be blamed for the tree's success; rather, it was federal biologists and other turn-of-last-century promoters who got the expansion going.[13]

At that time the public never envisaged its spread as a threat. By the 1950s, when a quarter of a million Chinese tallow trees were reportedly growing in and around Houston, nobody raised an alarm. Residents admired the attractive poplarlike tree that thrived in their grass-filled, swampy city. Tallow prospers in wet, exposed habitats, where the climate is mild and freezes are light and infrequent. It is flood tolerant and grows well on rivers, lakeshores, floodplains, and marshes. The Gulf Coastal Plain and Blackland Prairies ecoregion fitted it perfectly. People increasingly selected it as a yard tree and planted it along highways. It was hard not to like in the landscape. The tallow's encroachment on old fields and abandoned acreages was viewed simply as a boon. Even as late as 1960, Texas tree authority Robert Vines reported Chinese tallow was only "sometimes escaping."[14] The tree simply did not appear ominous; it was more or less well behaved.[15]

From Boon to Bane

Starting around 1970 habitual recognition and acceptance began to change as the tree proliferated on upland prairie habitats, notably southwest of Houston. Biologists expressed concern about this explosive encroachment. In Galveston County, for instance, woodlands of tallow accounted for only 2 percent of the county's total land area in 1970. Ten years later encroachment had quadrupled and thereafter doubled every decade, accounting for 16 percent by 1990 and a (projected) 31 percent by 2000. Neighboring counties witnessed a similar unexpected expansion.[16]

Experts advanced reasons to account for its sudden movement. The large-scale abandonment of farms and ranches to make way for urban development provided disturbed space for the tree. Ample seed production, which wild birds consumed, also helped its spread. And there is the simple fact that lag times of several years between a plant's establishment and its explosive growth are not uncommon. Whatever the trigger, Chinese tallow became the first tree to successfully invade the Texas coastal prairie. Within a decade after arrival tallow shades out native grasses, such as little bluestem and Indian grass, and forms a wood. Not long after that, one model shows, tallow turns into a forest consisting of a single, invasive tree.[17]

Chinese tallow made inroads into a variety of forest types, including bottomland hardwoods, pines, and mixed woodlands. Tallow can be found in Texas' Big Thicket, for example, in mature forests around New Orleans, and on barrier islands near Charleston, South Carolina. It is the most successful exotic plant invader of native woodlands in southwestern Louisiana.

There are economic consequences to its success. The southern US forest, which stretches across 13 states, supplies 60 percent of the timber produced in the United States and is one of the most productive in the world. Tallow grows on more than 457,000 acres of this forested land, where researchers argue it is "the most pervading, forest-stand-replacing, alien tree species in the region."[18]

In 1996 the US Nature Conservancy designated Chinese tallow as one of the country's 12 worst "environmental scoundrels."[19] The Invasive Species Specialist Group of the International Union for the Conservation of Nature (IUCN) includes it as one of the world's worst invasive alien species. Today, the tree is considered invasive in the Gulf Coast states, as well as in Oklahoma, Arkansas, Tennessee, Georgia, South Carolina, and North Carolina. According to current distribution maps, the tree grows in one-third of Texas' counties, in 90 percent of Florida's, and in all parishes of Louisiana and has even made inroads into counties in California.[20]

Chinese tallow invading a mixed woodland cleared for a subdivision. Photo by Rebekah D. Wallace, University of Georgia, Bugwood.org.

Why Is the Tree So Successful?

As is the case with almost all exotics that become invasive, a number of factors account for the species' immense success. In the case of Chinese tallow, quirks of geography, climate, peculiar growth habits; freedom from native predators; and economic promise all converged to create an invasive dynamo.

In regard to geography, *Triadica sebifera* grows in every province of China south of latitude 30° north. The same parallel crosses the US Eastern Seaboard at Jacksonville, Florida (30° 19′), heads west through Mobile, Alabama (30° 40′), Biloxi, Mississippi (30° 24′), and New Orleans (29° 58′), before cutting into Texas close to Houston (29° 45′) and Austin (30° 18′). It is no surprise that the area along this meridian has the largest-sized stands of tallow in the United States. There, moist, warm conditions with mild winters are an ideal home for the exotic tree. The tallow would likely extend across the entire Southwest if it were not for increasing aridity, typified by an average of only 9 inches of rainfall in El Paso but 60 inches in Beaumont.

But the success of the tallow owes much to its habits. It is a stunningly fast grower. Given direct sunlight, Chinese tallow saplings can attain heights of 9 feet within just two years of germinating. Foot-tall saplings have soared to 13 feet in less than two years in south Florida. Even in shadier, forest conditions, as was noted in a river floodplain in southeastern Texas, saplings grow about nine inches a year, which is between 2 and 10 times the rate for native trees. Shade tolerance allows the tree to take hold in areas already forested; they can even establish themselves under closed canopies.[21]

Tallow trees start producing seeds at about three years of age and remain active producers throughout most of their approximately 50-year life span. Five-year-old trees can average a pound of seeds annually; 20-year-olds average more than 25 pounds annually. Yields from the early Bureau of Plant Industry plantations exceeded 10,000 pounds—or five tons—per acre. Seeds soaked in water for 30 days boast higher germination rates than dry seeds, which helps explain how tallows growing along estuaries and rivers led to the invasion of nearby prairies. It also suggests that the trees are naturally adapted to flooded and moderately saline conditions along the coast.

In addition to prolific seed production, Chinese tallow has other reproductive advantages. Root sprouting has been noted as far as 16 feet from the tree trunk, and subsurface rootstock may live longer than the trees themselves. Resprouting is common after trees are cut or burned. Tallow roots are not picky about soil types—clay, loam or sand, acidic to slightly basic—and the trees' own leaf litter imbues those soils with nitrogen concentrations (i.e., fertilizer) nearly double what is found in soils growing native grasses.

But, as is the case with practically all truly invasive species, the real key to the tree's success lies in simply being free of the predators in its ancestral home. In China, *T. sebifera* is subject to at least 26 species of plant-eating insects and several species of disease-causing bacteria. Naturally, most, if not all of these, are absent in the United States, and it will likely take some time before American insects can evolve their digestive systems to tolerate the tree's leaf chemistry. Some studies suggest that tallow in the Texas coastal prairie exhibits only 1 percent leaf damage from insects, compared with 26 percent for native green ash and as much as 15 percent for willow and hackberry.[22]

Freed from munching insects, the tree no longer has to manufacture chemical defenses; it can pour resources into growth and reproduction. The average total mass, including shoots and roots, of an invasive tallow tree in Texas is significantly greater than that of a tallow tree in China. According to a hypothesis known in ecological circles as the evolution of increased competitive ability, the natural conditions of the new or nonnative environment lead to actual genetic shifts in growth traits. Several studies have shown that even when American strains of tallow are grown in China, they outperform their Chinese counterparts. Despite being readily attacked by local insects and diseases, these reintroduced American trees are able to compensate for the loss by growing prolifically. These invasive strains of tallow from the States even display altered physiological characteristics that allow for greater soil nitrogen uptake. In the space of a century or so, natural selection has created a new super type of Chinese tallow that is adapted to thrive in a new and basically competitor-free environment.[23]

Scary Trends: Birds, Hogs, and Weather

Other factors are helping Chinese tallow thrive. Native bird species consume tallow's fatty seeds, which, studies have shown, germinate more readily when they have passed through the avian gut. Crows are apparently so fond of the seeds that one of the common names for the tree in China honors the bird. To date, 35 US wild birds have been observed feeding and foraging in tallow stands. Some species, such as yellow-rumped warblers, American robins, northern cardinals, and northern flickers, feed heavily on seeds. Yellow-rumps and Baltimore orioles take special interest in the seeds on the Texas coast; the former scrape off and consume the waxy coat, while the latter swallow the seeds whole. White-winged doves in South Texas descend on the trees in large numbers, plucking some seeds whole while shaking others to the ground. Most authorities agree that the foraging habits of these birds affect tallow abundance, at least locally. To what extent birds are a major factor in the regional spread of the tree is less certain. However, the remarkable northward spread of these doves during the last three decades has likely enhanced the tree's abundance in urban and suburban locations inhabited by the birds.[24]

Any single change to an ecosystem has multiple reverberations. On the one hand, Chinese tallow's presence offers a new food source. Its seeds are energy sources for several winter resident birds, such as yellow-rumped warblers. The yellow rumps are unusual among North American warblers because they are physiologically capable of assimilating tallow's high-melting-point fatty acids. They have the unique ability to switch from feeding on insects in conifer and deciduous woodlands where they nest during the summer, to taking fruits of the native bayberry and wax myrtle during the winter. Both the bayberry and myrtle are chemically similar to tallow. Experts hypothesize that Chinese tallow may actually be influencing the winter distribution of the warblers, whose populations have increased recently in tandem with tallow's dominance on the upper Texas coast, for example, and in some urban areas.[25]

On the other hand, other bird species may not be benefiting from the tallow feast. Just because they consume the seeds does not mean they are absorbing valuable energy. Research shows that northern cardinals, important dispersers of tallow in Louisiana, fall into this category. One also has to consider what the birds would otherwise be eating if they had something other than tallow at their disposal. Migrant and wintering birds require high-energy, easily assimilated carbohydrates, such as the sugars found in flesh fruits. Tallow supplies a lipid-rich food for wintering birds and nourishes only those that can handle the special fats it offers. More important, the near absence of insects in tallow, especially moth and butterfly caterpillars, deprives spring migrants of energy-intensive foodstuffs after their long Gulf crossings. For these northbound birds, a dense tallow woodland might as well be a desert.[26]

A related problem is that the abundance of tallow potentially deprives native plants of their natural dispersal agents. The native flora that offer fall and winter fruits with some lipid value, such as flowering dogwood, yaupon holly, and several species of hackberry and greenbrier, have lower lipid levels than tallow. Research suggests that even a mild preference for tallow over native plants could mean the natives are outcompeted as dispersal agents.[27]

Adding insult to injury, feral hogs are also helping tallow trees spread more readily, at least according to a seven-year study in the Texas Big Thicket. Researchers found that tallow trees were twice as abundant in areas where hogs were rooting than in areas where they were excluded. The abundance might be the result of elevated soil nitrogen due to the hogs' disturbance of surface litter and/or their dispersal of seeds throughout foraging areas. Regardless, the notion that one exotic mammal is helping an exotic tree is a cruel twist in the invasive story.[28]

But the most ominous news about tallow and its status as an invasive species simply relates to climate. The two main climatic constraints on the tree are low temperatures and low precipitation. Even with these constraints, climate models using current conditions predict the tree has the potential to spread more than 300 miles northward beyond its current boundary into the US Midwest and Mid-Atlantic states and includes possible naturalization in the Pacific Northwest. Areas where water is scarce, such as the southern Great Plains and Great Basin, still have extensive rivers and reservoirs that make for suitable habitat. Current climate models already project that tallow will spread into the drainage basins of 21 of the 23 major rivers in Texas.

And when scientists attempt to factor in a warming climate, the situation gets worse. Allowing for a 3.6°F increase in daily minimum and maximum temperatures, climate models show tallow trees extending more than 430 miles northward beyond their current distribution. This means as far as Ohio, Pennsylvania, and on the coastline of southern New England.[29]

Increases in hurricanes and flooding (whether fresh or salt water) will not impede the advance of the tree, which is more adapted to inundation than many native species. In fact, the surge from Hurricane Katrina, which became trapped for three weeks behind levees in a hardwood forest near New Orleans, only *increased* the dominance of tallow over less flood-tolerant eastern persimmon, sugar hackberry, possumhaw, and red maple. Chinese tallow can easily withstand four to six months of immersion during winter and as much as one month during its growing season. At the Three Gorges Dam, constructed on the Yangtze River in China in the heart of tallow country, the tree was one of only a handful of hardwoods that could tolerate a half-year's winter submergence in the reservoir's drawdown zone—the area between winter's high- and summer's low-water marks—which is 100 feet in vertical height.[30]

Is Management Possible?

What can be done about the tallow? Completely eradicating the invasive tree—at this late date—is impossible, or at least economically unfeasible. But doing nothing also has costs. One study estimates that, in the absence of management or control, tallow will occupy nearly three million extra acres within 20 years. Estimated losses of native timber alone will run up a price tag of $300 million.[31]

The only real options are burning and mechanical removal. Natural prairie fires once limited woody growth in grasslands. Not surprisingly, fire used for plant control works best when it mimics natural processes in repeated and frequent use. Such use essentially requires unpopulated natural areas that are fairly dry and that have sufficient herbaceous fuel to burn. A prairie with plenty of grass and forbs would be perfect for fire control of tallow, as long as the trees are mere saplings. Once mature, the trees do not provide the fuel for an effective fire. What paradox is this? Wood everywhere but nothing to burn?

Part of the problem is that Chinese tallow is well adapted to fire. The thicker bark of mature trees protects inner layers from injury so that even hot fires fail to kill the trunk and branches. After a blaze, the tree responds by vigorously resprouting from the base of its trunk (with growth up to six feet in a year), or from its roots, which helps *spread* the tree clonally many feet away. Really hot fires can ignite tallow, but in general the trees do not propagate crown fires.

But the real problem with tallow—the truth of the paradox—is that stands of the tree competitively exclude the understory grasses and shrubs that are needed to drive an effective burn. "It is common to watch a prescribed fire burn right up to the edge of a tallow stand and simply go out because of a lack of fuel," says Jim Grace of the National Wetlands Research Center in Lafayette, Louisiana. In effect, mature stands of tallow "render the ecosystem nonflammable."[32]

For fire to be a practical management tool, one has to burn the trees when they are mostly saplings under six feet in height and when there is still sufficient fuel present to burn them completely. Beyond this size, the trees require repeated burns, mowing, and applications of herbicide to control resprouting. In short, as is the case with so many invasive species, catching tallow at the *beginning* of a local outbreak is critical.

Mechanical removal is useful where fire is not practical, such as in urban areas or on farmland and ranches, along canals, or among isolated pockets of trees. Bulldozing and disking the soil will work tolerably well for farmers and ranchers, but several years of intensive maintenance afterward is necessary. The use of shredding mowers, followed by a dense application of the resultant tree mulch, also helps. Tallow seeds germinate best in soils with fluctuating daily temperatures; reducing fluctuation by using mulch discourages germination. Still, herbicides will have to be applied to the tree trunks, and the seedlings that do emerge have to be cut or pulled. A study in Florida found that seeds were still sprouting in herbicide-treated plots five years later. Mechanical removal is labor intensive and expensive.[33]

Unlike the examples for water hyacinth and hydrilla, to date no biological control agent exists for Chinese tallow, though there is hope on the horizon. The flea beetle (*Bikasha collaris*), a common herbivore that attacks this tree in south-central China, is exhibiting traits that experts find promising in quarantine studies: high specificity for the target weed and an inability to complete its life cycle on other closely related plants. The beetle larvae feed on tallow roots, while the tiny (less than one-tenth inch) adults, devour the leaves. If approved for field release, the flea beetle may contribute a much-needed tool for tallow control.[34]

In short, management of tallow woodlands currently requires repeat burns and/or mechanical removal. All situations demand monitoring and managing over several years, given the tree's root sprouting and prolific seed production. One thing everyone agrees on is that early detection and control yield the best outcome and land managers should prioritize the early invasion of new areas over mature stands.

Tallow as a Resource

Beekeepers, however, are strong advocates for the tree. One study from southwestern Louisiana in the 1970s concluded that Chinese tallow was one of the top-10 most important nectar sources for honey production. For about six weeks in the late spring the trees yield an amount of nectar rarely equaled by most trees in the United States. Experiments with honeybees in the Houston area noted that 400 colonies produced a remarkable 30,000 pounds of honey from approximately 300 acres of tallow forest. The high-quality honey is amber in color and has a "mild though excellent flavor."[35] Beekeepers like the fact that each tallow tree blooms for about two weeks, and tallow woodlands account for a continuous, six-week bloom beginning in early May. One bee enthusiast proclaims Chinese tallow "the most successful tree nectar source ever introduced into the United States."[36] Beekeepers have even opposed legislative attempts to add the tree to lists of noxious weeds. Given the recent bee colony collapse crisis, it may be politically difficult to do anything that could be construed as stressing bees further.[37]

Chinese tallow wood, fiber, and leaves offer a few, though little-exploited, options for use. Tallow wood is a fuel, of course, and when properly dried it burns well. Plantations in the Houston area have produced more than four tons of oven-dried wood per acre per year with little agricultural assistance. Arguably, burning wood for electricity production is kinder to the environment than burning fossil fuels. Tallow fiber, mixed with bagasse (matted cellulose fiber from sugarcane processing), has good potential for medium-density fiberboard; such use spares ever-more-scarce and more expensive woods. Tallow leaf litter, rich in nitrogen and tannins, could provide fertilizer and other industrial uses.

And what about those waxy seeds, which sparked the tree's importation in the first place? The saturated fat of the tallow on the seed's surface is edible, or at least can be processed to be edible. The Chinese have created a transfat-free shortening from these fatty acids. But it is the oil extracted from the kernels of the seed, or stillingia oil, that has the most potential to grab headlines in the States. As an unsaturated fatty acid, it can easily yield a substitute for petroleum.[38]

Petroleum Substitute?

Since 1957 the USDA has been searching for local oil-rich plants that can reduce our dependence on petrochemicals and thereby on imports of foreign oil. Fuels from organic material, such as biodiesel, are typically more environmentally friendly, emitting lower amounts of carbon, sulfur, and particulate matter. The main problem in the widespread manufacturing of biodiesel is the high cost of feedstock oils, which are usually refined oils. Using less-expensive, nonedible oils would alleviate that problem.

Chinese tallow is a promising candidate for a biodiesel. First, with a nod to China and Ben Franklin, the tree is a powerhouse oil producer. "The Chinese Tallow tree must be regarded as one of the most productive oil-seeds of the world, if not the most productive," proclaim authorities on the tree.[39] When the tree is cultivated, annual oil yields run to 4,000–5,000 pounds per acre, roughly half in the form of tallow and half in stillingia oil. In liquid terms, this is somewhere between 645 and 970 gallons per acre. This yield is "more than any other plant species now growing in the US," notes an economic botanist.[40] With recent research indicating that stillingia oil biodiesel can be used in engines without any modification, it would seem a potential energy boom is waiting in the dense tallow woodlands.[41]

David Whetsell, a retired businessman in South Carolina, formed Whetsell Energy to harvest tallow tree seeds out of the wild to make such biodiesel, as well as high-protein feed for poultry and livestock. Moreover, working with Clemson University and the University of Georgia, Whetsell is learning how to *cultivate* the tree for maximum production. His company aims to plant millions of acres in tallow trees in the US South, to reduce our foreign dependence on oil, provide jobs, and build the economy.

Wait. *Cultivate an invasive species?*

The idea harks back to the hippo enthusiasts who suggested seeding nonnative water hyacinths to increase forage for a nonnative beast. Cultivating Chinese tallow, however, does not seem quite so far-fetched. Arguably, the tree is not "wild" to begin with. It has been "selected, improved and cultivated by traditional Chinese methods for possibly a thousand years," experts remind us.[42] In this light, a tallow woodland is arguably more like a cornfield than a forest. Even better than a cornfield, tallow woodlands require no supplemental inputs, are relatively free of insect damage, and grow in marshy and wet areas, so-called unproductive or marginal lands not suited for agriculture, the same ones that the hippo enthusiasts pointed to. That these lands are not arable is important; it means tallow crops do not have to compete with our foodstuffs, as is the case with soybeans and corn, which, when grown for biodiesel, vie with the same products destined for human and livestock consumption. Finally, being a tree crop, tallow needs planting only once, unlike annual crops.

Rising sea level will benefit Chinese tallow. The tree's remarkable tolerance of poorly drained areas helps researchers conclude that tallow "may be the ideal energy crop for biodiesel production along the Gulf Coast,"[43] producing "much more oil per acre on wetlands than any conventional temperate-zone oilseed can produce on good farmland."[44]

Two big issues arise here, one practical and the other philosophical. On the practical side is the feasibility of cultivation. If the past 250 years of tallow growth in the United States has taught us anything, it is that there is a wide gap between projections and reality, between a possible resource and an actual product. Just vary the seed yields or oil composition a bit, and the economics can quickly head into the red. Harvesting raises concerns, too. Can the process be mechanized efficiently, and can the uneven maturation of the seeds be dealt with effectively? Will coppicing be required? In addition to experts at Clemson and the University of Georgia, other researchers at institutions such as Louisiana State University are experimenting with different harvesting methods.

On the philosophical side, is it appropriate to cultivate a nonnative invasive species that many would prefer to eradicate? Given the environmental and economic costs of this tree, can we really encourage more growth? After all, the tree has never been successfully cultivated to any large extent anywhere outside China, and its history in this nation is rife with failure.

This must have been the reaction of the Texas Department of Agriculture when confronted by Dennis Fisher, CEO of the now-defunct, California-based company BioCentric Energy. Fisher's company had developed a new microwave technology for oil extraction, and, rather than harvest tallow seeds, it found that the entire tree could be ground up, the hydrocarbons extracted, and a biodiesel produced after simple refining. With a microwave facility located in Orange County, Texas, the company was considering planting Chinese tallow in rows and shredding the whole lot every two years. With the plant being one of the few trees on its invasive species list, The Texas Department of Agriculture said "no way," not without a variance.[45] Fisher turned to contractors who cut down tallow trees along the highways. He not only managed to obtain a healthy supply of the trees but even was paid to take them off their hands. Despite the company's failure—a cash-flow issue—the Orange County venture suggests there might be an economically feasible avenue to control and perhaps eradicate the tallow woodlands.

In the end, the "most useful plant" that Franklin introduced to America in 1772 never realized the potential he envisaged for it. It has supplied no marketable candles, soaps, cosmetics, plastics, varnishes, floor waxes, or fertilizers. While biodiesel fuel still leaves room for hope, Chinese tallow is about dreams and intentions. However, usefulness is always contextual—to someone, somewhere, at some point in time. Outside that context, there are no guarantees. Instead of marketable waxes and oils, we are left with vanishing prairies, disturbed forest and timberland, birdseed of some value, and more valuable amounts of honey. Chinese tallow is a cogent reminder that what is useful is always relative and unpredictable.

The lesson of Chinese tallow goes beyond material gain. Aesthetics and practicality play their roles, but so does the perception of harm, which is at the heart of what an invasive species is. Gulf Coast residents admired an easy-to-grow tree that was beautiful year-round. Families assembled under its shade, benefiting from the tree's spread branches on an otherwise open prairie. Tallow comforted people and livestock and never posed the threat to commerce, human health, or recreation that water hyacinth and hydrilla did. It was only after the trees began an explosive expansion into coastal prairies, recognized today as one of the most threatened ecoregions in the state, that Texans began to express concern, but fitfully. It is hard to understand harm to an ecosystem until our pocketbooks or health is under assault. Accustomed to planting trees around our homes for privacy and shelter, it is difficult for us to admit that some lands are better left treeless. Chinese tallow is a telling reminder of how we literally do not see the forest for the trees.

5 The Risk of Tamarisk
The Saga of Saltcedar

Saltcedar touches on a host of issues concerning invasive organisms. Its initial appeal and reason for introduction were purely aesthetic, but over time its utility as shade, erosion control, and a windbreak accounted for its spread across the nation and state. What led to invasive stands and thickets, however, was alterations we made to the rivers in the American West. Just as reservoirs have helped water hyacinth and hydrilla, changes to river hydrology created a niche around dams that saltcedar (or actually a hybrid between two species) has been able to exploit. The tree, unlike Chinese tallow, proved to be useful. It simply worked too well, outcompeting native trees, further altering already-changed ecosystems and even sheltering an endangered animal, which has complicated biocontrol efforts. It is hard to praise or blame saltcedar. It reminds us that we simply do not know how an exotic organism will always behave, especially in disturbed watersheds of our own creation.

Saltcedar near mouth of Santa Elena Canyon, Big Bend National Park. Photo by Matt Turner.

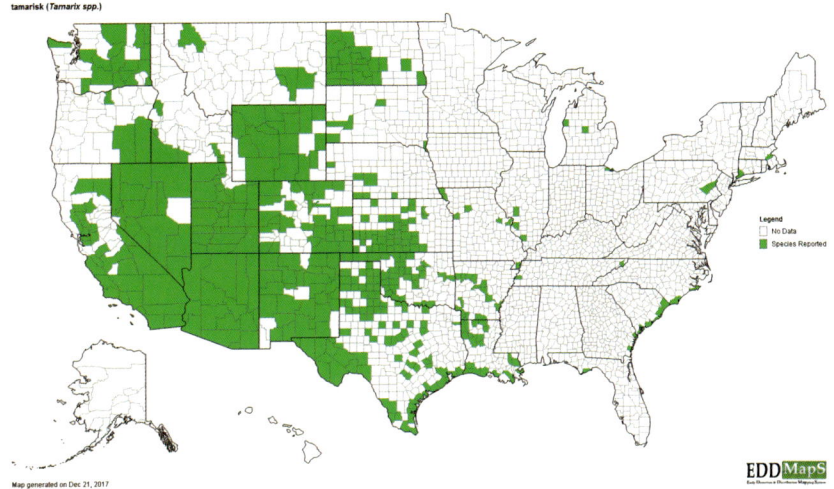

tamarisk (*Tamarix* spp.)

Legend
☐ No Data
◼ Species Reported

Map generated on Dec 21, 2017

EDDMapS

US distribution of saltcedar (*Tamarix* spp.). EDDMapS, University of Georgia—Center for Invasive Species and Ecosystem Health, 2017, http://www.eddmaps.org/.

Few Americans had more influence on the modern environmental movement than Aldo Leopold. From his days as a forester in the 1920s through his role as a professor of game management in the 1930s and a US presidential adviser, Leopold was a champion of regarding humankind and the natural world holistically. Leopold found value in preserving wilderness for its own sake, and he regarded the scientific approach to wildlife as a way to restore balance to existing ecosystems. He expounded on these ideas in a posthumously published book, *A Sand County Almanac*. Considered a cornerstone of conservation science and philosophy, the slim work calls for and defines a "land ethic" that binds animals, plants, water, and soils with humans into an indivisible community. One of the book's best-known passages is a succinct summary of this ethic: "A thing is right when it tends to preserve the integrity, stability, and beauty of the biotic community. It is wrong when it tends otherwise."[1]

How astonishing it is in hindsight to learn that in about 1920 the author of these words actually planted a tamarisk tree, more commonly known as a "saltcedar," in front of his house in Albuquerque, New Mexico. Not only is the saltcedar alien to the flora of the United States, but it has proved to be "one of the most successful nonnative plant species in western North America."[2] Especially dominant in desert riparian ecosystems, the saltcedar has ended up spreading into every perennial drainage of the arid US Southwest, forming pure stands along many of the region's major rivers, including the Colorado, Gila, Pecos, and the one near Leopold's house, the Rio

Grande. Arguably, no tree could have been more counter to Leopold's land ethic than the one he planted. Saltcedar has altered stream morphology, increased wildfire frequency, soaked up precious water, and—far from preserving integrity—has just plain *taken over.*[3]

In Leopold's defense, the saltcedar was, in the words of his hydrologist son Luna, "rather uncommon" at the time, and the full development of his father's land ethic was still several decades away.[4] Aldo could not have foreseen the invasion that was to come; and this was his front yard after all, not a huge national park or massive unspoiled wilderness. But the incident touches on the issues so common with invasive species, saltcedar in particular. Is it ever good to import an exotic species? Do we fully anticipate how it will behave in its new home? Can we figure out all the costs and admit to some benefits about its introduction? Finally, if a land ethic implicitly includes humans in the land community, what are our responsibilities and limits as good stewards of the environment? Is a land ethic only for native species or for other ones as well? In light of these questions, some experts have begun to soften their condemnation of the tree's derring-do ways.

Origins and Hybrids

The genus *Tamarix* includes 40 to 50 species and hybrids of mostly shrubs and trees that carry numerous slender branches and small, scalelike leaves. The pale pink to white flowers that blossom on these woody plants are tiny and arranged in showy plumes that are attractive. Seven of the eight species currently in the United States and Canada are deciduous large shrubs and small trees (reaching about 20 feet high). Like most tamarisk species, they are superficially similar, making them difficult to distinguish, even by experts. The eighth species, the athel tree, is distinct, an evergreen growing up to 60 feet tall.

All these species and hybrids are commonly called saltcedars. As might be expected, no member of the genus *Tamarix* originates in the New World. The native distribution of the genus stretches from North Africa and the Mediterranean basin eastward to North India and west-central China. The genus's center of distribution probably lies in the dry valleys of the Middle East, extending as far as Pakistan, but many centuries of use prevent us from delineating its geographical origin more precisely.

The saltcedar is well known in the Old World. The tree is mentioned several times in the Bible's Old Testament. Reportedly, Abraham planted a saltcedar at Beersheba ("the well of the oath") to honor a covenant he made (Genesis 21:33). Saul, the first king of Israel, addressed his followers while seated under a saltcedar (1 Samuel 22:6). He was later buried under a salt-cedar tree after he famously killed himself by falling on his sword (1 Samuel 31:13). Some scholars have suggested that *Tamarix gallica* is one of many candidates for manna mentioned in the Bible. Saltcedar manna is created when a boring insect penetrates the stem and causes the formation of a gall, which exudes a "sweet, white, somewhat nutritious sap."[5] Cultural and historical familiarity with saltcedar probably contributed to the tree's welcome in the New World.

The two species considered to be the most invasive in the United States and that make up large, dense stands in the western states are *T. chinensis* and *T. ramosissima*, or their hybrids. *T. chinensis* is native to eastern China, and the more widespread *T. ramosissima* hails from north-central China across to eastern Turkey. Remarkably, DNA evidence suggests that 85 percent of the saltcedar invasion in the United States consists of hybrids between the two. In other words, by planting the two species together, we have unintentionally encouraged genetic combinations missing in the Old World. These novel genotypes may help explain the plant's dynamic expansion in the United States.[6]

First Contact and Early Years

Generally, the saltcedar was respected from ancient times as one of the few trees that could provide shade and greenery in a torrid zone. Ornamental feathery branches and pinkish blooms led people to propagate it. It is believed the Moors brought saltcedar with them into the Iberian Peninsula and thence to western European gardens. Precise details are missing, but saltcedars likely crossed the Atlantic Ocean to the States from Europe or the United Kingdom in the early 1800s. In 1818, mention is made of a French saltcedar (*T. gallica*) growing in the Botanic Gardens at Cambridge, Massachusetts. Nursery catalogs on the Atlantic coast sold saltcedars as early as 1823, and 30 years later they were being advertised on the Pacific Coast. Various species were introduced after 1800 both as ornamental specimens and curiosities.[7]

Arboriculturists appreciated its growth habits and uses. In 1868 the USDA housed six species in its arboretum in Washington, DC, noting that their shrubby forms and salt tolerance—hence the name saltcedar—suited them for seashores and dunes. Perhaps introduced to prevent sand erosion, the first report of a naturalized stand of saltcedars came from Texas, where a doctor visiting Galveston in 1877 described them "growing on the island in the greatest profusion, along ditches, the borders of ponds, etc."[8] By 1886, the US Army Corps of Engineers in Texas was planting the same shrubs at

Saltcedar was first promoted in the United States in the early 1800s as an ornamental. This is a particularly showy specimen from a yard in Santa Fe, New Mexico. Photo by Matt Turner.

the southern end of a barrier island called San José (where Hurricane Harvey first made landfall in 2017) to stop sand from clogging the dredged channel that led from the Gulf of Mexico into Aransas Bay.[9]

Saltcedar's reputation as an agent to control windblown sand and topsoil spread. And any ornamental shrub that could flourish in heat and drought without irrigation was bound to attract attention, especially in the US Southwest. One government staffer working for the USDA's Section of Seed and Plant Introduction called saltcedar "the most drought-resistant and otherwise hardy of all the trees and shrubs" planted on his Texas Panhandle farmstead; and "there appears to be no limit in dryness of the soil on any usual Great Plains' farm beyond which this plant will not survive."[10] Land

and water managers began to promote it as a shade tree and windbreak. It prevented soil erosion and stabilized the banks of streams and reservoirs. Ranchers began inserting switches of saltcedar along riverbanks. While experts debate to what extent the tree was actually planted and to what extent it spread by itself, by 1900 saltcedars flourished from Texas westward as far as Southern California. It was likely in this atmosphere of excited curiosity and appreciation for useful, climate-appropriate plants that Leopold set his saltcedar in his Albuquerque yard.

The Catalyst: Damming the West

The real catalyst for a saltcedar "explosion" came not from nurserymen, soil managers, ranchers, or curious scientists like Leopold but rather from the US Congress. The Reclamation Act, which President Theodore Roosevelt signed into law in 1902, allowed monies from the sale of semiarid public lands in 16 western states and territories to be set aside for massive irrigation projects. These projects would, in the words of Frederick Newell, the first director of the Reclamation Service (initially within the US Geological Survey [USGS]), turn the "dead and profitless deserts" into arable land for farmers and homesteaders.[11]

In the first five years the fledgling service took on 30 reclamation projects. Then, under its more robust and independent role as a bureau within the US Department of the Interior (in 1907), the agency set about making dry areas look green. The bureau spearheaded the construction of more than 600 dams, restricting the flow of nearly every major river in the American West. Over the following six decades three-quarters of the country's 100 largest reservoirs were contained by some of the highest and largest dams in the world.

Despite shortcomings, the Reclamation Act produced beneficial results, at least initially. The West became more extensively inhabited and developed. More than 10 million acres of farmland, including 60 percent of the nation's vegetables, are still watered under the act's irrigation mandate, marking the region, notably California's Imperial Valley, as one of the premier croplands in the world. Dams' hydro plants generate some 40 billion kilowatt hours of electricity annually, enough to power 3.5 million homes. Most of our larger southwestern cities such as Phoenix, Tucson, Salt Lake City, Las Vegas, and Los Angeles could not exist in their current size if this act had not been passed.[12]

The Reclamation Act watered the West, but it also fundamentally altered the hydrology of the entire region and inadvertently created the perfect conditions for a saltcedar takeover. In the mountains and deserts, most prolonged periods of high water happen during the spring from the melt of winter snowpack. The ensuing floods used to leave behind large stretches of bare, moist sandbars and scoured riverbanks, perfect for native riparian trees, such as willow and cottonwood, to disperse seeds onto exposed sediments. Their seeds are only briefly viable: the period just after the high water in spring is their main opportunity to sprout.

Reservoir managers, however, want to hold back floodwaters, irrigate crops with the extra flow, and distribute it throughout the summer months. The switch in water use exposes the receptive sediment not in the spring but in early fall. By then, willow and cottonwood have finished seeding, which allows saltcedar, producing copious amounts of wind-borne seeds all summer long (up to 600,000 seeds per mature tree), to germinate on any spot of exposed earth with little competition. Moreover, as reservoirs are lowered, they expose "bathtub rings" of damp silt, which saltcedar all too readily colonizes.

Damming rivers in the American West opened a niche for saltcedar, but what allowed the plant to exploit that niche so completely are its roots. The tree is a phreatophyte, which means it obtains a large portion of its water from the phreatic zone, or water table. Saltcedar can send down taproots to greater depths than any native plant competitor. Cottonwoods and willows are also phreatophytes, but they have relatively shallow roots that access groundwater 7–10 feet belowground, whereas a saltcedar can push down its taproot twice as far. A saltcedar also possesses large numbers of fine roots that soak up moisture throughout the soil profile. The tree can also spread out from multiple root-crown sprouts.

Saltcedar's deep and complex network of roots helps it thrive under an amazing variety of conditions. Having originated in desert areas, the plant handles drought, and in riparian habitats, its deep roots allow it to grow away from the river channel onto floodplain terraces that typically have no native riparian vegetation. At the wet extreme, a saltcedar can tolerate being submerged for as long as 70 days and partially submerged for almost 100. And as the water table drops, the roots drop with it, so saltcedars flourish in highly variable conditions. These same conditions tend to keep native species in check. This unique tolerance makes saltcedar "the most successful exotic tree in the western United States."[13]

Expansion and the Shift of Opinion

Between 1880 and 1920, saltcedar spread rapidly along the river systems of the US Southwest. It grew in the Great Basin and in watersheds of the

Rocky Mountains. Historical photographs tagged the rate of spread to be about 12 miles per year. Although experts debate to what extent the tree had already spread prior to the river damming, few disagree that our manipulation of river hydrology strongly favored saltcedar. And at the very moment when people might have entertained second thoughts about its value, additional reasons surfaced to keep planting it. The 1930s was the decade of the Dust Bowl, when 100 million acres of Great Plains topsoil literally blew away because of drought and poor agricultural practices. It is hard to fault federal land managers and officials for turning to the saltcedar as a way of anchoring whatever soil was left.[14]

But opinion did change. Erosion control and windbreaks were a godsend, but junglelike thickets across river bottomlands were not. As saltcedar continued its march, the very characteristics that made it seem so useful began to turn it into a monster. Experts complained it was expanding too far and too fast. Its multifunctional root system stymied erosion but encouraged silting and sedimentation and soaked up too much water. Dense shade created by too many saltcedars discouraged the growth of native plants, reducing a diverse riparian landscape to impenetrable thickets of a single species.

By 1920, saltcedar covered approximately 10,000 acres in the United States. By 1960, its coverage spanned 900,000 acres. Almost a third of it was in the Pecos River basin of New Mexico and Texas, where the tree blanketed the entire Pecos River bottomland from the New Mexico border south to a point below Sheffield, Texas, 175 miles away. While that area held a huge number of trees, saltcedar was popping up in other locations around the state. Trees already a foot in diameter were growing in the Panhandle in 1914, and by 1948 one of the state's foremost botanists, Lloyd Shinners, declared the tree was likely to be found "in every part of the state."[15]

There is no doubt the saltcedar invasion had serious consequences. The nonnative plants exacerbated the hydrological changes that humans had wrought on watersheds in the West. Prior to dams, river bottoms consisted of unstable sand deposits that shifted and moved during floods. River channels were wide and shallow, often with multiple low islands and sandbars. Banks were flexible, changing shape with the volatile flows but rarely overflowing except during the strongest floods. Under natural conditions, native plants might temporarily gain a foothold along the banks and bars, only to be scoured out by the next heavy rainfall. This maintained a flat and broad riverbed.

After we regulated river flows, however, the contours of the beds began to change. After reservoirs were constructed, discharge volumes were cur-

Saltcedar grows profusely along both sides of the Rio Grande west of Lajitas, Texas. Looking to the south or Mexican side, the trees appear as a dark green band with fuzzy blooms growing just behind the bankside vegetation. On the US side, the invasive giant reed (foreground center) appears with saltcedar (at right). Photo by Matt Turner.

tailed and the usual scouring ended. Beds and islands did not shift, and riverbanks, channels, and bars grew fixed and inflexible. Saltcedars, having already gained a foothold, aggravated the situation. Their many small stems and extensive roots created drag and slowed flow rates even further, increasing sediment buildup and reducing any scouring. The alien trees stabilized riverbanks and sand deposits, solidifying channel features. Midstream islands that were once three feet above the water surface tripled in height, and the wide, shallow channels grew narrow and deep. The channel width of the upper Brazos River in Texas, for example, was reduced by 100 yards after saltcedar began to fix its banks. Ironically, narrower channels lead to overbank flooding, even with minor discharges. While experts debate whether river regulation or saltcedar invasion is more to blame for riparian changes, most concede that the two were tied together.[16]

Another feedback loop involves saltcedar and wildfire. Saltcedar's very fine and dense leaf structure is highly flammable, even when trees are green and well watered. And the leaf litter that collects in the many branches and on the ground during winter adds fuel to the fire (literally), making trees easy to ignite even when bare. While fire in native riparian woods is relatively rare—usually occurring only in severe drought—it is relatively frequent in saltcedar-invaded areas, with regular fire intervals of 10 to 20 years becoming practically the norm. While cottonwoods, willows, and saltcedar all resprout after fire damage, the former two (both natives) recover poorly as fire intensity increases. Saltcedar, however, with deep roots and root crowns under the soil, can tolerate fire, both in frequency and intensity. The cycle becomes apparent: as saltcedars spread, fires increase; and as fires increase, saltcedars spread even farther. This fire cycle, detrimental to native species, encourages saltcedar dominance and expansion.[17]

Other downsides of saltcedars, such as their reputation for causing soil salinity and for guzzling water, turn out to be less clear cut. Saltcedars have salt glands in their leaves that remove and secrete salts that the trees pull from the subsoil. The salts from leaf litter accumulate on the ground, and many scientists argue this prevents the germination of other species' seeds and helps explain why saltcedar thickets tend to be monospecific. Experts are divided on the extent to which this really happens. Some concede there are higher salinity levels in saltcedar-dominated areas but point to other factors that mitigate the salinity, such as surface evaporation in arid climes or overbank flooding.[18]

Saltcedar's reputation as a water hog has not withstood prolonged investigation and close scrutiny. What really galled experts in the 1940s and 1950s was the belief that saltcedar guzzled water. Water is scarce in the Southwest, and clearly something so widespread was doubtless taking more than its due.

In 1939, USGS hydrologists estimated that completely killing river-bottom vegetation (much of it saltcedar) would liberate 70,000 acre-feet of water annually. In 1942 a USDA irrigation engineer estimated that there were 50,000 acres of saltcedar in the Pecos River basin in southern New Mexico and that each acre of trees guzzled five acre-feet of water per year (about the amount 20 households in the desert Southwest would currently use in a year).[19] Put another way, one researcher estimated the value of water lost to saltcedar in the western United States in 1998 to be between $133 million and $285 million annually.[20]

Measuring water availability and use is a complex matter, however, especially under field conditions. It basically turns out that saltcedar's water use was, understandably and unintentionally, being exaggerated. Two important reports in 2009, one by the USGS and another by the Tamarisk Coalition, a nonprofit organization founded to restore riparian corridors, discovered that a saltcedar consumes roughly the same amount of water as does a native streamside plant, such as a cottonwood or a willow. The reports carefully distinguished expected water savings (by removing salt-cedar) from actual water availability. As others have noted, the "prospects for salvaging large quantities of water for human use by removing tamarisk are probably illusory, as 50 years of failed demonstration projects show."[21]

Whether soil salinity and water guzzling were real or partially conjectured, the general consensus after about 1950 was that there was simply too much saltcedar and something had to be done. Saltcedar had become the most common riparian shrub or tree in most river systems in Southern California, Nevada, Arizona, Utah, New Mexico, Oklahoma, and Texas. By the late 1960s overall estimates of saltcedar coverage approached 1.3 million acres. Saltcedar eventually began to make the noxious weed lists of states outside the Southwest, including Colorado, Wyoming, Nebraska, North and South Dakota, Oregon, and Washington. Beyond the United States, saltcedar has become naturalized in Argentina, Australia, Canada, Mexico, and South Africa.

Ongoing droughts in the American Southwest, with their water restrictions, golf-course closures, and alarmingly low reservoirs, increased the pressure to do something about the saltcedar. It was in this light that Congress passed the Salt Cedar and Russian Olive Control Demonstration Act, which President George W. Bush signed on October 11, 2006.[22] Commentators noted that the act "was the only significant invasive-species legislation to pass in that session of Congress, and with nearly unanimous bipartisan support."[23] Recently, bipartisan interest in resolving an invasive plant problem is rare. The Salt Cedar Act required the Department of the Interior to assess the extent and impact of the invasion of the two plant species; provided funding of $80 million for large-scale demonstrations and research projects; and targeted long-term management and funding strategies. Unfortunately, with the financial crisis of 2007–8, the ensuing Great Recession, and foreign wars, only a small portion of the funding became available.

Management, Control, and Beetles

A variety of techniques aim to get rid of or control saltcedar. Two methods, doomed from the outset, have been burning and flooding. As one might guess, prescribed burns do not work since the trees are adapted to fire. And saltcedar's tolerance for all types of water conditions makes flooding ineffective. Mechanical removal is equally vexing. The problem is that if one removes the aboveground trunks, branches, and stems, a tree will rebound. Mechanical removal essentially acts like fire: it encourages regeneration from the roots. In fact, removal of top growth stimulates vigorous, denser growth than was originally present, resulting in a shorter, bushier canopy. Saltcedar crowns and lateral roots can be excavated, but this increases labor costs and causes excessive soil disturbance.

Chemical treatments have had good results. However, riparian areas are sensitive to chemical runoff and water pollution. To date, only a few products are deemed environmentally safe for saltcedar removal: 2,4-D, glyphosate, triclopyr, and imazapyr. Alone, these products are not particularly successful, but in combination they can kill 99 percent of the trees. Naturally, the chemicals kill any vegetation they contact, alien or native, and should not be sprayed along perennial streams. Imazapyr, a nonselective herbicide, is expensive and remains toxic in soils for up to a year, curtailing the regrowth of native plants. Nonetheless, it has received a federal rangeland- and aquatic-use designation. Helicopter spraying with a herbicide containing imazapyr along 289 miles (13,497 acres) of the Pecos River and tributaries brought about a saltcedar mortality of 80–90 percent over a six-year period (1999–2005).[24]

The newest tool in saltcedar control, and the one that has stirred the most excitement, is a tiny beetle (one-fifth-inch long). On its native turf, 325 species of insects and mites munch on the tree. Selecting one for biocontrol

Tamarisk beetle (*Diorhabda elongata*), one of several species used as a biocontrol agent against saltcedar. Photo by James L. Tracy, US Department of Agriculture, Agricultural Research Service.

Saltcedar damage by tamarisk beetle larvae. Photo by Eric Coombs, Oregon Department of Agriculture, Bugwood.org.

has not been easy. Robert Pemberton of the USDA's Agricultural Research Service (ARS) singled out a beetle that feeds on the trees in Mongolia. Pemberton and other colleagues started a saltcedar biological control program in the 1970s. Jack DeLoach, also with ARS, followed up in the 1980s. *Diorhabda elongata deserticola*, as the beetle species was then named scientifically, showed promise. Imported beetles were held in quarantine at facilities in Temple, Texas, starting in the early 1990s, and in Albany, California, by the end of that decade.

There's nothing complicated about how the *Diorhabda* beetles control saltcedar; they simply devour its leaves. More than 80 percent of the damage occurs during three caterpillar-like larval stages, with the remainder left for the adult stage when they actually change into beetles. Their entire life cycle takes 30 to 40 days, which allows the beetles two or three cycles in a typical growing season. Their foraging completely defoliates the saltcedar. The tree responds by flushing with new leaves, which the beetles also consume. One loss of foliage does little damage, but repeated feastings on replacement foliage starve the tree of precious carbon. Unable to maintain its basic metabolism, the tree slows down leaf production and growth and in some cases dies. Seedlings and younger trees are the most susceptible to beetle infestation.

Mature trees take several years of repeated leaf losses to succumb; therefore, it may take a decade or more for a saltcedar stand to disappear.[25]

After a surprise stumbling block (explained later), in 1998 the USDA's Animal and Plant Health Inspection Service (APHIS) gave a go-ahead for field-cage trials with *Diorhabda* beetles at 10 sites in Texas and six western states. Considered benign for the ecosystem as a whole, the beetles were officially released at seven sites in 2001. Results were promising, but the first real saltcedar biocontrol success story was on a site near Lovelock, Nevada. By the close of 2002, Lovelock's beetles had devoured 2.5 acres of trees; a year later more than 700 acres, and a year after that a whopping 25,000 acres. Since then, beetles have become established at numerous sites across the US Southwest. In Texas, more than 800,000 beetles have been released across 15 counties in western parts of the state. Experts report defoliated saltcedar thickets along miles of the Rio Grande and Pecos, Colorado, and upper Brazos Rivers.[26]

Success has not been uniform, however, and lots of tweaking was needed to match the beetle with saltcedar habitat. Careful taxonomic study determined that what had been called *D. elongata* was actually a cluster of five distinct species. Three of them are now successfully established in North America. The northern tamarisk beetle (*D. carinulata*), originally from northwestern China and Kazakhstan, is the most widespread, inhabiting the intermountain regions of eight western states. The Mediterranean tamarisk beetle (*D. elongata*) lives in multiple sites in Texas and central California. The subtropical tamarisk beetle (*D. sublineata*), originally from Tunisia, is the latest introduction. Released in 2009 along the lower Rio Grande in Texas, it has defoliated more than 60 miles of saltcedar.

Most experts agree that the beetles are doing a good job but will not eradicate the invasive plant. Most effective results combine all these methods. For instance, land managers can mechanically clear mature trees, then use prescribed burns or herbicides to suppress regrowth. Or they can spray herbicide and follow up with burns or beetles. They can even release beetles first to weaken trees and then burn them. The advantage of using fire, at some point, is that the flames consume all the woody material left over from other controls. Clearing felled trees can cost more than the actual method of control.

Finally, refiguring the conditions that resulted in invasion is an obvious place to start. One way is to restart the spring flood regime. Restoring this natural occurrence is unlikely to extirpate saltcedar, but it will promote the appearance of native plants. When cottonwood and willow are given a chance, they outcompete saltcedar. In fact, saltcedars are only a minor problem in unaltered streams where flood pulses occur. And along altered rivers that allow periods of extensive soil waterlogging, such as the middle third of the Rio Grande in New Mexico or in the Colorado River delta in Mexico, saltcedar coexists with native species in a patchwork mosaic.

An Unexpected Find

What almost brought beetle biocontrol to its knees in the first place, and what has since inhibited its recent use, was an unassuming bird. Dull gray and barely six inches long, the drab willow flycatcher (*Empidonax traillii*), which migrates from Latin America to nest in the United States, has been declining in much of its western range. The subspecies that summers from California to West Texas, and aptly called the southwestern willow flycatcher, was listed as endangered in 1995. Like its kin, it prefers thick foliage along rivers, swamps, and wetlands. In the US Southwest this habitat consists of willow, cottonwood, seepwillow, box elder, and buttonbush; however, saltcedar has taken over much of that riverine habitat—likely dealing a blow to the flycatcher.

The federally endangered southwestern willow flycatcher. Photo by Jim Rorabaugh, USFWS.

Bird experts were surprised and delighted to discover that, far from shunning the saltcedar, the little flycatcher has turned to using the invasive tree for its nest sites. This discovery happened in 1994, when the bird was being proposed for listing as a threatened species and just when beetle researchers were asking APHIS for permission to test their saltcedar-killing beetles. Charged with threatened-species oversight, the US Fish and Wildlife Service expressed concerns about losing saltcedar. APHIS conducted new studies that suggested the beetles would benefit or at least be benign for the flycatcher.[27]

As a compromise, APHIS agreed to release beetles at least 200 miles from any known southwestern willow flycatcher nest sites. They took additional precautions: researchers had to use secure cage trails, then carefully monitor releases on the sites, and, only then, move for a general release. Monitors insisted they had conducted "one of the most extensive and complete risk analyses yet attempted for any biological control program."[28]

Despite all the precautions, tamarisk beetles spread more quickly than expected, so in 2010 APHIS terminated the biocontrol program. Although no longer officially released, these leaf eaters are still spreading and have gotten into the nesting range of the flycatcher. From releases in Utah, the insects have flown south into northern Arizona and New Mexico, and there is at least one instance of them gaining access to flycatcher habitat in the Virgin River watershed that runs into Utah, Arizona, and Nevada. Initial worries on the part of the US Fish and Wildlife Service seem to have been warranted.[29]

The bird or beetle dilemma raises two important points. The first is, whether we like it or not, saltcedar plays a role in disturbed riverine environments and provides ecological services there. The tree is a nonnative species that helps prevent bank erosion, slows land-surface runoff, emits oxygen, and provides shade and shelter for wildlife. The flycatcher has been able—in the face of radically altered water regimes and vegetation type and coverage—to take advantage of a saltcedar-dominated landscape. It is a fact to marvel and wonder at. When people looked more closely, they found no fewer than 50 bird species using saltcedars, and the number increases when saltcedar is mixed in with native trees. One can argue, ironically, that the invasive saltcedar has been a saving grace for the flycatcher and possibly additional birds. USGS experts argue that the alien plant permits birds "to persist in their historic range and in some cases to spread in areas where they would not occur otherwise." They also comment, "In some areas tamarisk may be the only option for a functioning riparian forest."[30] Clearly, there are second thoughts about the suitability of having the tree where it is.

Trends in saltcedar management are focusing more on restoring riparian areas than simply killing off the foreign trees. In some areas, completely eradicating the tree would do more harm than good because it does not result in the regeneration of native trees, except through replanting. In fact, the habitat may have been so transformed and disturbed that completely restoring its original character is unrealistic. Some biologists also believe, as we have alluded to in this book, that we have gotten too carried away with a native-versus-nonnative formula, calling the former good and the latter bad. We need to move beyond this simplistic distinction, they say, and consider the entire spectrum of roles nonnative plants, notably saltcedar, play in any ecosystem. This nuanced position regards saltcedar perhaps as neutral or even beneficial in certain situations. It reflects a change more in emphasis than in substance. To most experts, saltcedar is still to be removed, even on a large scale, but doing so involves care and forethought and having Leopold's admonition about keeping the biotic community healthy in mind.[31]

The other point that the birds-versus-beetles issue raises is about biocontrol. Can we really ever know what will happen when we set free an exotic organism? Biocontrol enthusiasts are optimists. They tell us that of the 350 control organisms released in 51 nations over the past 150 years, only 8 attacked nontarget plants, and not one caused serious economic or environmental damage. Furthermore, in North America biocontrol has been employed against weed species about 40 times since 1945, resulting in partial to full control of two-thirds of the target species.[32]

Beetle researchers studied saltcedar control for 20 years and took ample precautions before they freed the insects. They have achieved remarkable success. But could this success go as far as to "change western riparian areas as much as did the invasion of [saltcedar] itself"?[33] Would the extirpation of saltcedar lead to the extinction of the southwestern willow flycatcher? We do not have answers for such theoretical questions. We can speculate. Perhaps the beetles can bring some kind of equilibrium, whereby they lower saltcedar densities enough to give native trees the opportunity to rebound, but not too low if the flycatcher is to survive. Perhaps the beetle will eventually begin eating native plants, as grass carp have done with aquatic vegetation in Lake Austin. Nobody can say with certainty. The experts do their best, as efforts to control hydrilla show. In addition, there is always pressure to find a solution, a quick fix, or a problem that can be "solved" during this or that funding cycle.

Management and control involve a time line that may be historically too short or too arbitrary in respect to what we have in mind. Attention and commitment to a project may be as much personal, social, and cultural as it is ecological. Experts with the US Geological Survey have stated an important truth by declaring any biocontrol agent to be "a large-scale ecological experiment with unknown but potentially widespread consequences."[34] We must keep in mind the lesson concerning the arrival, spread, and ongoing management of saltcedar because it involves complex factors that are time and place dependent.

We carried this storied tree to our shores, promoted it, allowed it to hybridize, and even created new habitat that it was able to exploit. Then we turned on it and brought in exotic organisms to stymy its abundance and spread. Saltcedar reminds us that nature is a dynamic process whereby organisms adapt to all kinds of change. It also reminds us that we are agents of profound environmental change, and each change leads to new changes that we cannot always anticipate. We consider invasive species as organisms imposed on us, but we are more and more aware of our responsibilities for importing and releasing them. Saltcedar holds up the mirror in which we see ourselves. It shows the extent of indelible changes we have made in the American Southwest and causes us to redefine our role of land stewards going forward.

6 Feral Cats
Kitties and Killers

Threats that sparrows and starlings pose to cavity-nesting birds or that beetles pose to flycatcher nest sites in saltcedar pale when compared with the destruction caused by cats to backyard birds and additional animals. Many urban residents believe that wild tabbies they encounter in woods, parks, and alleys are lost or abandoned pets. Some are, but many are also wild born. Organizations are seeking to nurture these cats, characterizing them as companions. Many members and others argue that, unlike hogs raised for meat, domesticated felines deserve humane treatment. Other citizens, however, regard cats as instinctive predators that are capable of decimating wild animal populations. People take moral stances about feral cats, whether it involves suffering to the cats themselves or to the numerous animal species on which they prey.

No invasive species has been the topic of such an emotional roller coaster as the domestic cat. The tabbies that roam downtown alleys, suburban neighborhoods, vacant lots, and city parks have opponents and advocates, who challenge data, trade insults, and even threaten physical violence. Opponents want the feline carnivores—abandoned, stray, wild, or partially roaming (words vary)—to be trapped, gassed, poisoned, or shot because of the damage done to native animal populations. Advocates want the same cats to be fed, protected, and treated well. They want the public to care for and hopefully adopt free-roaming kittens and adults (if not too wild) and urge city officials and animal volunteers to manage the well-being of the felines in residential areas.[1]

A major player in what has become an acrimonious standoff is the Maryland-based Alley Cat Allies, self-styled "national voice for cats," claiming 650,000 supporters. Alley Cat Allies reported almost $7 million in assets in 2016, devoted to the humane treatment of cats, especially outdoor felines. It opposes the trapping and euthanizing of the feral animals, working to saving cats' lives in shelters and promoting state laws and city ordinances to protect all cats.[2]

Alley Cat Allies and similar groups with strong commitment to humane treatment actively promote a program called Trap-Neuter-Return (TNR), whereby feral cats are live-trapped, checked medically, vaccinated, surgically rendered sterile, and then returned to their capture sites, with the tip of the left ear removed to show they have been neutered. Hundreds of community-based animal groups, municipal agencies, and organizations for the ethical treatment of animals support TNR. They regard it as an effective way for allowing feral cats to live, mostly in the areas where they have been born, while reducing populations over time through neutering.

Alley Cat Allies and similar advocates are opposed to bird-watcher groups, including the National Audubon Society, the American Bird Conservancy (ABC), and other wild animal organizations such as the Nature Conservancy, which insist TNR neither reduces the overall number of feral cats nor lowers predation on local wild animals. Opponents insist TNR does not wean cats of their instincts and habits of killing native wildlife. Cats are instinctive predators, they insist, so by prolonging the lives of feral cats, cat lovers are increasing the number and the suffering of birds, frogs, insects, and other animals killed by cats.

This is not always the case. A report about New York's Neighborhood Cats, a group that offers practical advice about TNR care for a colony (often a group of related animal individuals), notes that volunteers, who carefully manage their feral cats, do see a reduction in felines. It appears that well-managed colonies in which *all* members are neutered, a policy New York City has switched to, can lower numbers among the city's 2,000 reported feral cat colonies. However, neutering all members in a colony is hard to do, especially when the number of individuals is large and may grow even larger after more and more cats (together with raccoons and opossums with disease issues) are drawn to feeding stations.[3]

Writing a blog for the American Birding Association, Brian Monk, cat owner, bird lover, and veterinarian, takes an opposing view: "TNR can neither eliminate feral cats, nor reduce predation, and does not address illness or disease, facts supported by actual scientific study." Monk accuses cat advocates of acting on sentiment rather than reason, of ignoring and distorting facts and research, and resorting to personal name-calling with people who disagree with their views. Based on his professional expertise, Monk states, "TNR does not ease feline suffering or eliminate feline predation on our wildlife to a point that is acceptable, to me as a veterinarian and a conservationist, or to anyone else who considers the facts." He concludes that trapping followed by euthanasia is a viable solution.[4]

Cat advocates, however, claim that TNR with food and shelter provided by human caregivers depresses the need for cats to destroy wildlife. Besides, cat supporters continue, killing cats is more financially expensive than neutering and returning them and merely creates a momentary vacuum that neighboring wild felines, which average five kittens per year, fill. With an estimated 80 million cats out there, there are just too many animals to kill.[5]

Domestication

Keeping cats goes back at least 9,500 years, confirmed by a grave site in Cyprus excavated in 2004 that contained a skeleton of a cat and presumably its human caregiver. The skeleton resembles the African wildcat that genetic specialists argue is the wild progenitor of today's common tabby. Domestication happened in the Middle East during the time of early agriculture, though DNA evidence suggests that tamed cats continued to intermix with wild kin for millennia. Unlike the situation with dogs, there is very little genetic distinction between modern domestic and wild cats.[6] The Romans reportedly spread domesticated cats throughout their empire, carrying them into Britain about AD 300, where the Scottish wildcat, an isolated population of the European wildcat (*Felis silvestris*), is being genetically swamped through breeding with domestic cats. Today, that endemic feline is the most endangered species in the British Isles, with perhaps 35 purebred wild cats at large.

The discovery and settlement of North America resulted in domestic cats becoming popular during colonial times—killing stowaway rats and mice. Surviving rodents scurried ashore into houses and yards. Today, about one US household in three owns at least one cat, and that fraction is growing. According to the American Society for the Prevention of Cruelty to Animals (ASPCA), the population of owned cats is 85 million or higher with another 60–70 million additional stray or unconfined cats. Globally, the estimate exceeds 600 million for all cats, but nobody knows for certain.

The debate about stray or feral cats involves three major issues. The first is the predatory impact that cats in general, but especially feral cats, have on birds, small mammals (voles, shrews, and chipmunks, as well as unlovable invasive rats and mice), reptiles (snakes, skinks, and anoles), amphibians (frogs), and invertebrates (butterflies, dragonflies, and worms).[7] The second is about the health risks feral cats, and to some extent domestic ones, pose to human caretakers and other animals. The third is about the ethical responsibilities we must consider in regard to catching, neutering, releasing, feeding, or managing these long-standing animal companions while also seeking to protect many native species on which cats prey.

The cat's keen ability to hunt rodents helped bring about a close relationship between humans and felines over millennia. Photo by Lxowle (CC BY-SA 3.0).

No other invasive animal has been as revered as a companion and also condemned as an arch predator on wildlife and a secondary threat to human health (passing on toxoplasmosis contained in its feces). Conflicting perceptions are at the heart of this battle. It depends on whether you regard cats as family members, aloof pets, work animals, nuisances, or pests. The iconography of perceiving them as gods or witches mirrors our ambivalence. Unlike traditional food and products supplied by a feral hog, the emotional connections we gain from a house cat or tending a wild one cause us to confront a dilemma—keeping cats while also valuing many animals on which they prey.

Problems Posed by Predation

Historical research suggests that cats have played a role in the extinction of at least 22 island bird species and in the declines in many more. The Global Invasive Species Database managed by the Invasive Species Specialist Group (established in 1994) with the IUCN's Species Survival Commission notes that feral cats kill endangered species and may alter the fauna of local ecosystems. In respect to New Zealand, where one in two households currently owns a cat, the database states, "Feral cats have been implicated in the decline of at least six species of island endemic birds . . . as well as 70 local populations of insular birds."[8] The international organization has placed the domestic cat among 14 mammals, including the goat, pig, red fox, and macaque monkey, on its list of 100 of the world's worst alien species. Once established, cats are hard to get rid of. They have been eradicated from fewer than 100 of the approximately 9,000 small and medium-sized islands around the world on which they have gained access.[9]

The cat's instinct to pounce on insects and small animals, especially birds, is perhaps the most important reason for people's dislike. Alley Cat Allies challenges this opinion, stating that humanly caused pollution, habitat loss, and the degradation of the environment due to urban and industrial growth take a much greater toll than feral cats do. The Alley Cat Allies website states that "sidestepping the issue of human destruction to focus on trivial but sensational issues, such as the so-called 'cat versus bird' debate, only diverts attention away from the enormous and far more dangerous impact of humans." This is a valid statement. Historically, we must claim responsibility for overhunting species to the point of extinction, such as the passenger pigeon and great auk. We must also take responsibility for introducing new predators, such as rats and cats, into areas where they did not exist and where native species have no defenses against them.[10]

The pro-cat group draws attention to bird deaths due to more recent factors, such as collisions with wires, towers, buildings, and vehicles, and to pesticides and the loss of habitat due to human population growth and industrial development. By situating predation by cats, whether having undergone TNR or not, within this larger perspective of negative agency on the part of human beings, Alley Cats makes a plausible point.

There is an array of variables associated with habitat transformation and loss due to urban and industrial expansion that are causing measurable declines in local bird and other wild animal populations. Well-known field and garden birds have shown dramatic decreases in North America and Europe during the last 30 years or so. Habitat destruction is clearly part of the reason. However, Alley Cat Allies prefers to downplay the impacts of unowned and free-ranging cats by reminding us about these structural issues while also stressing that cats have always done what cats do. Their conclusion is that we human beings must behave more intelligently and responsibly to conserve our native biota.

What Alley Cat Allies overlooks is that the importation and release of nonnative cats is itself a dangerous human impact on the environment. It fails to mention, say members of the bird fraternity, how the multiplication, spread, and concentration of felines, notably in and around urban areas, is another complicating factor, a very significant one, in wild animal loss and suffering. A briefing paper issued by Texas Parks and Wildlife's Conservation Outreach Program underscores the predation aspect and argues that TNR does not do the job that people claim. It regards these programs as ineffective. They "do not prevent overpopulation of feral cats, reduce population size over time, prevent losses to native wildlife, or prevent disease transmission," it concludes.[11]

The Humane Society of the United States, which promotes the *Neighborhood Cats TNR Handbook*, estimates that about one in five cats is adopted from an animal shelter, with about 860,000 euthanized every year.[12] Kittens are best candidates for placement in rescue and adoption programs, and this is one active aspect of TNR to socialize them to be house cats. The millions of feral cats, most of which are too wild to be adopted, live in densities that can be as high as hundreds per square mile in metropolitan areas—more than nature can support, claims California's Department of Fish and Game biologist Ron Jurek.[13]

Such densities reflect the animal's ability to adjust rapidly to changing conditions. "Cat populations are extraordinarily capable of reaching local carrying capacities as a function of reproductive mechanisms that emphasize breeding efficiency," note the authors of the TNR analysis.[14] Cats can adjust the size of home ranges, rates of population growth, and overall numbers in a given area. Part of this adaptability is due to opportunism.

A pride of feral cats. Photo by Sara Golemon (CC BY-SA 2.0).

They can hide under bird feeders, raid nest boxes, and stalk fledglings. They catch rats, mice, and lizards in all kinds of places; forage for scraps behind restaurants; and pick through dumpsters in modern high-tech urbanscapes. Cats are excellent scavengers as well as predators.

A big problem for many bird enthusiasts is not just that cats kill birds but that they kill so many of them. According to a systematic review of predation, domestic cats kill between 1.4 and 3.7 billion birds and between 6.9 and 20.7 billion mammals annually in the United States. Considering these range of estimates, authors Scott Loss, Tom Will, and Peter Marra conclude surprisingly, "Our findings are that free-ranging cats cause substantially greater wildlife mortality than previously thought and are likely the single greatest source of anthropogenic mortality for U.S. birds and mammals."[15] They included other sources of human-related mortality such as striking windows, buildings, towers, and moving vehicles, as well as ingesting pesticides.

Efforts to reduce predation depend on two things. One is to make outdoor cats as conspicuous as possible to a prey species. A recent promotion of so-called cat bibs in the form of brightly colored materials hung from the animal's collar is one way of doing this. The bib approach is a step forward compared with bells and other noise-making devices that are supposed to alert potential victims to a cat's presence. They do not work well. However, the decorative, visible bib that reportedly does not impede a cat's mobility is better suited to house cats than feral ones. And research suggests that feral cats take more animals than other felines.

Feral cats wreak havoc on local bird populations, especially in parts of the world where no similar natural predator exists. Free-ranging cats (feral or not) are the single greatest source of anthropogenic mortality of birds and mammals in the United States. Feral cat with bird (left) and rabbit (right). Photos by NottsExMiner and Eddy Van 3000, respectively (CC BY SA 2.0).

Another "solution" is to keep our tabbies inside the house and, whenever possible, to arrange for feral cats and their progeny to enjoy similar living spaces. Bird groups such as Audubon urge owners to keep their cats indoors. They argue that housebound animals live longer than feral counterparts and enjoy healthier lives. A range of toys, window seats, and enclosed porches can stimulate them. Many bird groups insist that the house pets enjoy a full life without having to go outside.

The ABC, which regards feral cats as a significant threat to bird conservation, agrees. ABC does not support TNR, believing it to be largely ineffectual, bad for both cats and birds. Its solution is to support local ordinances that seek to control cat behavior outdoors and to make it illegal for owners to abandon or leave cats unsupervised. Bird lovers prefer that owners tag their cats with microchips or, where necessary, register them and keep them from roaming. The ABC also asks the public to stop feeding outdoor cats. They point to trapping and taking them to shelters for possible adoption as indoor pets. However, many people are neither able nor willing to have feral cats inside their homes. Most groups would agree that keeping cats indoors helps both the felines and wildlife. However, this is not a realistic solution for most adult stray and feral cats, which are too wild to be kept indoors. Thus, a no-kill policy keeps feral cats outdoors, hopefully, but not necessarily, in better health than unsupervised wild felines.[16]

PETA also supports keeping cats inside a house or apartment. Claiming to be the largest animal rights group in the world with three million members and supporters, PETA prescribes toys, scratch posts, seats or cushions beside windows, and screened porches for keeping cats happy. It also recommends walking cats on a leash. PETA justifies its position, noting, "Feral cats rarely, if ever, die of old age." They catch communicable diseases, die from accidents, become laboratory animals, or are subjected to callous and brutal treatment. PETA states that feral cats live short, often painful lives. Accordingly, while the organization is not opposed to trapping and neutering, it does not in good conscience advocate TNR "as a humane way to deal with overpopulation." TNR is acceptable only under special circumstances, one of which is not having access to wildlife. PETA does not oppose euthanasia. This may be the most humane solution due to "the huge number of feral cats and the severe shortage of good homes, the difficulty of socialization, and the dangers lurking where most feral cats live," it continues. Therefore, "It may be necessary—and the most compassionate choice—to euthanize feral cats."[17]

Feral Cats and Health Issues

Cats contract illnesses and can pass some of them to human caretakers. Transmitted infections take place through contact with cat feces and the saliva from an infected individual. There are four kinds: viral, parasitic, bacterial, and fungal.

The viral disease rabies, which attacks the central nervous system, poses a deadly problem for a cat after an infected raccoon, bat, skunk, or fox—the most common carriers of rabies in the United States—has bitten it. Cats are very susceptible to this disease; thus, vaccinations are important to keep them healthy. Some local ordinances require cats to be vaccinated against rabies. Fortunately, no cases of cat-transmitted rabies to humans have been identified in the United States for many years. Other core vaccines include feline viral rhinotracheitis (FVR), feline calicivirus (FCV), and panleukopenia (FPV) dispensed as a single shot. FVR causes flulike upper respiratory infection that is very contagious to other cats. FCV has similar symptoms. FPV is a distemper-like parvovirus that can be fatal in cats.

Infection from a parasitic microbe can prove serious to both cats and humans, notably people with depressed immune systems. About one in five Americans carries *Toxoplasma gondii*, which becomes dormant after a few days of flulike symptoms. Humans get the virus mostly by consuming undercooked meat and unwashed fruit and vegetables. However, hunting cats feed on raw meat so may pick up the parasite, which completes its cycle only in these felines. Thus, an infected cat can shed millions of oocysts, infectious forms of the parasite, into water and soil, where they remain for months. Owners are encouraged to change a litter box daily (as the parasite is not infectious until one to five days after being excreted) and to wear gloves and wash hands after doing so. The cat is the only domestic animal that is a definitive life host for *Toxoplasma* and is a known vector for transmission to humans. Currently there is no vaccine against toxoplasmosis.[18]

Estimates suggest as many as 40 percent of the world's cats are infected with toxoplasmosis, with rates in the United States varying between 16 and 80 percent, depending on the locality. There may be upward of 3 billion people worldwide, including 60 million in the United States, also infected with toxoplasmosis, and some experts suggest that the parasite may cause behavioral changes, including personality shifts and other neurological disorders, such as schizophrenia. However, most people show no clinical signs of the disease.

How many humans contract toxoplasmosis from cats is not known; it is easier to become an accidental host from ingesting undercooked meat and contaminated water. Feral cats are more likely to carry the protozoan than house cats. One recent review suggests that between 9 and 46 percent of pet cats in the United States, Europe, and South America show a past exposure to the parasite. And food and water contaminated by cat feces, together with infected meat, notably pork, is the usual path for toxoplasmosis to be transmitted to people.[19]

Zoonotic diseases include bacterial infections resulting from scratches and bites that may result in swollen lymph nodes in the upper torso and possible fevers, headaches, sore muscles, and fatigue in people. Infections from so-called cat scratch fever (*Bartonella henselae*) cause about 25,000 cases in the United States annually. It is believed that as many as 40 percent of household cats carry those bacteria at some stage of their lives.

Exchanges of intestinal parasites, such as tapeworms and hookworms, are also possible, especially when children play in sandboxes and soils contaminated by cat feces. People touching a cat's infected fur may also transmit the fungal infection known as ringworm.[20]

Morality and Compromise

Is it right for us to revile an animal for being what it is, a natural predator? Cats instinctively hunt rodents, birds, reptiles, and insects. We praise them for killing pestiferous rats and mice but condemn them for killing songbirds and other animals we value. This is our problem and not the cat's. Keeping cats indoors, as many bird enthusiasts suggest, may help an individual feline to live longer while suppressing or diverting its urge to hunt live prey. You can have your cat and enjoy wild birds provided you keep both of them as far apart for as long as possible. That is the mantra of the bird groups.

This may be true for house cats, but not for wild ones. Most feral cats are too old, wild, or temperamentally ill-suited for adoption and life indoors. And that is a problem given their ability to breed quickly, exist in large numbers, and depend on native prey as food. Cats often live in places preferred by birds and bird lovers, who set out feeders for garden species. Cats are agile and resourceful, adept at using cover to stalk and capture prey. Suburban gardens, neighborhood parks, and greenbelts are excellent habitats for predatory cats. However, fighting over territory and the diseases associated with crowding and stress reduce a wild cat's longevity. Both cat lovers and cat opponents recognize that feral cats do not live as long as household counterparts. Some advocates provide food, shelter, and living space to improve the quality of their lives. They are pleased to have local cat colonies or groups living in areas where animals were born.

Opponents recognize these issues as well but are less sanguine about the

By all accounts, feral cats lead shorter lives than house cats and can suffer from a variety of ills, like this stray near Corpus Christi, Texas. Photo by Terry Ross (CC BY-SA 2.0).

lasting benefits of voluntary support and whether adding food and shelter reduces predation on native species. Data that have been generated about the humaneness of TNR and its effectiveness in reducing cat numbers or slowing down predation by feral cats continue to occasion argument, not consensus. A Public Library of Science (PLOS) survey found that most care-takers do not admit their cats reduce bird populations or transmit diseases. The majority of bird conservers vehemently disagree and oppose any com-promise apart from euthanasia.

Rather than a lack of data, it is disagreement over what those data mean that polarizes opinions, note the survey's authors, who speculate whether a geographical approach might build consensus. They suggest that keeping feral cats away from highly sensitive sites, such as bird-nesting colonies, for example, could be agreed to. In return feral colony managers could operate in lower-priority or less sensitive areas. They note that wildlife conservation has shifted from valuing animals because they are useful to us to regarding them as integral to a particular ecology, aesthetically important, and mor-ally defensible as having intrinsic value. Accordingly, perhaps the predation issue can be treated with the same innovation and imagination that mark this larger shift in attitudes. If so, it could lead to genuine consultation and compromise. Continuing to accentuate the pernicious effects that cats have on wildlife in an increasingly technologically crowded and disturbed world, which is inevitably whittling down many wild animal species, does the opposite. Such negativity discourages compromise and simply hardens and reinforces positions.[21]

There is common ground. Both sides recognize there are too many feral cats. Both suggest that neutering is a worthy method of control, and most also agree that TNR is not going to solve the population problem. Euthanasia is a major sticking point in this regard. The ASPCA states that 40 percent of the 3.4 million cats and 30 percent of the 3.9 million dogs that enter animal shelters every year are euthanized. Five times more dogs are returned to their owners than cats, although slightly more cats (37 percent) are adopted than dogs (35 percent).

The city of Austin, Texas, prides itself as the largest no-kill city in the United States. This means that fewer than 10 percent of animals in shelters are euthanized. So when the shelter is overcrowded, cats are passed through the TNR process and returned to the property they came from, whether residents want them or not. This is hard for a socialized cat that ends up on the street and has been dependent on humans for food and shelter. It may be equally disorienting for a feral animal that has been highly stressed and turned loose again. The no-kill policy that has gained such momentum and advocacy ends up not being as humane as rescue groups suggest.

Veterinarian Paul Barrows offers an insightful suggestion. Having pondered the issue firsthand, Barrows regards the feral cat as basically an educational problem. Cat owners need to understand how to keep a cat responsibly and set about doing so. Most dog owners adopt guidelines, such as using a collar labeled with the owner's name, exercising on a leash, and not permitting animals to foul and stray; and so should cat owners, he insists. For example, responsible ownership means their pet does not become an ecological threat, a reservoir of disease, and a social nuisance by being abandoned or allowed to roam freely and reproduce at will.

In the latter sense, Barrows regards feral cats as a form of biological littering; and similar to earlier problems surrounding waste, it has become politically and socially inappropriate to litter. The same can be done for feral cats. Funds expended for TNR, feeding, and then reabandonment would be better directed toward owner education, Barrow continues. Moreover, animal-control ordinances should be implemented in respect to issues covering nuisance or harm, and feral cat caregivers must fulfill custodial agreements with city authorities.[22]

This is one nod toward listening to and thinking about the positions taken by the groups most polarized about free-roaming cats. Margaret Slater, who taught in Texas A&M's College of Veterinary Medicine and Biomedical Sciences, College Station, before joining the ASPCA, remarked that doing nothing is the worst solution. "Effective solutions can be developed only by finding common ground and bringing the strengths of each group to bear [on] the problem," she insists. This means interested parties need to sit down and work out flexible, site-specific solutions that effectively reduce cat populations, so that both cats and their prey may benefit. Given the divisiveness and stridency attached to the feral cat issue, there is a long way to go.

Is it possible to shift our premise from thinking about cats as unnatural because we have introduced them to the places they currently inhabit? We may not bear the responsibility for cats being what they are but do bear the responsibility about where they are. The places to which we are referring are mostly urban habitats inhabited by a Noah's Ark of native and nonnative species. So one option is to accept the cat into the new urban ecology we have unwittingly fashioned.

In a sense, the feral cat is a hybrid. It is a "wild" agile predator in an area we have come to think about as "tame." Cat advocates deemphasize wildness and treat it as a companion in a domesticated environment. They want to nurture and show compassion for it as an unwanted or lost pet, not as a predatory carnivore. Cat opponents emphasize the feral cat's wildness; it can be very destructive in and around urban areas, they argue, which are homes for both native and introduced animals. Cats prey on resident and migratory species that occupy or pass through urban ecosystems. So opponents argue that the feral cat is and should be subjected to pressures that predators face, such as diseases, accidents, and being taken by other carnivores (such as urban coyotes and possibly foxes and great-horned owls).

We have different and competing attitudes toward the feral cat. But there is one attitude we do share in common. It is to treat the cat as an opportunity for reflecting about the ways in which we honor life, nurture compassion, and exercise choice. The challenge the cat represents is about how we can accept and treat life in all its diversity and abundance.[23]

7 Red Imported Fire Ant and Friends
The New Plague

Ants are extremely adept at getting places—carried on soils, rocks, vegetation, earth-moving equipment, and other vehicles. This allows them uninvited access to new sites around the world. Experts know relatively little about half of the 14,000 or so species but have focused on a group of so-called tramp ants that have been shifted around the globe and do a lot of damage. They include five species listed among the world's worst invaders. Unfortunately, Texas has become a home for invasive ants, all of which arrived accidentally. One of the best known, the red imported fire ant, is the poster animal for how quickly and easily small insects arrive, become naturalized, and spread and how much environmental and economic harm they cause. However, Texas researchers are in the forefront of devising ingenious strategies for combating the red imported fire ant. They have raised and dispersed natural agents that disrupt ant colonies and help control populations. Additional ant species, however, have arrived and pose new challenges.

Students call it the Brack Tract, shorthand for Brackenridge Field Laboratory in the School of Natural Sciences, about three miles from the main campus of the University of Texas in Austin. Legend has it that in the not-so-distant past a mountain lion and her cub denned in the secluded area's thick brush and then swam back and forth across the lake. Well used, 1960s-era buildings and huts along the shores of Lady Bird Lake make the lab nothing special to look at. The 82-acre compound for plant and animal research is part of an unpretentious older Austin, a relic of native prairie and riparian woodland surrounded by the encroaching development of a burgeoning city.

But looks are deceiving. The field lab, where the mountain lion swam, not only assesses and stores data about biological change but is also the home of a nationally recognized insect collection and a focal point for invasive species research. Larry Gilbert, professor of integrative biology and the lab's director, reminds us that the Brack Tract "is the ecological equivalent to a cancer research center in biomedicine." Why? Because invasive species undermine the health and well-being of existing natural ecosystems,

The red imported fire ant worker, specimen from Brackenridge Field Laboratory, Austin. Photo by Alex Wild, University of Texas at Austin's Insects Unlocked project (CC0 1.0).

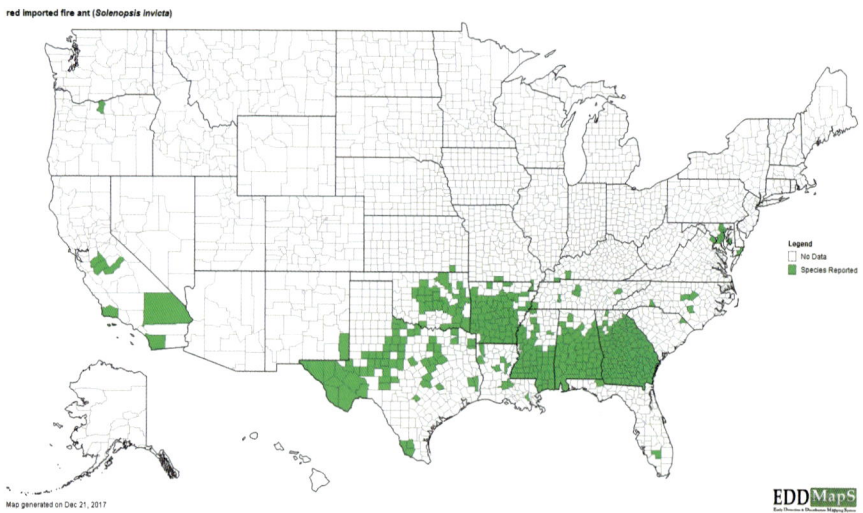

US distribution of red imported fire ant (*Solenopsis invicta*). EDDMapS, University of Georgia—Center for Invasive Species and Ecosystem Health, 2017, http://www.eddmaps.org/.

he explains. Gilbert and colleagues, supported by a multiyear grant from the Lee and Romana Bass Foundation, are investigating the control of a major pest with a natural predator that reminds one of a zombie movie: a brain-eating larva that devours the insides of its host ant's head in order to hatch into an adult that flies off to pursue new victims.

Public Enemy Number One

We tolerate this seemingly gruesome behavior because the victim in this case is an invasive insect, the tiny but pernicious red imported fire ant (*Solenopsis invicta* [RIFA]). During the past 60 years this aggressive ant, which occupies a swath of middle South America 2,000 miles long by 220 miles wide, has spread from its arrival in Alabama across the entire US South. Today, it infests at least 260 million acres of crops, pastures, parks, golf courses, industrial parks, and suburban neighborhoods in 14 states and Puerto Rico, and it is causing an estimated $6.7 billion worth of damage every year, including more than $1 billion in Texas. Many farmers and ranchers in the Lone Star State regard the RIFA, whose sting is painful and even deadly, as public enemy number one.[1]

RIFAs are aggressive, mobile omnivores capable of tolerating cold and

During flood events, fire ants can persist for days in floating masses. Photo by TheCoz (CC BY-SA 4.0).

A single fire ant mound can reach three feet across and, along with a labyrinth of subterranean tunnels, contain more than a quarter million ants. Photo by Robin Doughty.

heat, drought and flood. Mats consisting of large clusters of these ants reportedly floated with the consistency of gelatin and flowed like ketchup in Houston's flood after Hurricane Harvey in August 2017. Ant expert David Hu at Georgia Tech characterized these buoyant colonies as resembling a "waterproof fabric," capable of drifting for days and threatening anything or anyone who touched them.[2]

They are voracious predators on several agricultural pests that infest sugarcane, rice, soybean, fruit, and cotton crops. So in this respect they are beneficial and may also prey on ticks that latch on to field workers. However, the invasive species tunnels into crops, strips tree buds, and eats or damages the seeds and seedlings of at least 57 known cultivated species.

A mature RIFA mound is dome shaped and stands a foot or more high and as much as three feet wide, below which an underground labyrinth of tunnels may extend four or five feet downward. Each mound is home for upward of 250,000 ants, with a queen laying an average 1,000 eggs per day during her seven-year life span. Studies suggest that to support such numbers, RIFAs have to be exceptional competitors for food. Scouts rapidly locate potential supplies, and then foragers waiting in tunnels surface and set out to procure them. The foraging fire ants consume plant seeds, native fire ants and other insects, rodents, lizards, turtles, snakes, and the young of ground-nesting birds and mammals. The ants aggressively bite, then sting anything that crosses their path or disturbs them, including field workers and outdoor hikers. Gripping skin with their mandibles, they inject venom from their stingers. They reduce the amount of food, such as honeydew produced by aphids, whiteflies, and mealy bugs, favored by native ants. They also kill pollinator insects and are drawn to infest electrical outlets in houses and buildings, causing short circuits and the threat of fire.

The RIFA is life threatening to persons allergic to insect bites and stings. Allergic reaction to fire ant venom includes a high heart rate, difficulty in breathing, and swelling of the throat. In July 2001, an elderly man went into shock and died after being bitten in a nursing home in Georgia; his family received an almost $2 million settlement after it was disclosed that the facility had been infested by RIFAs for several years. In July 2006, an elderly woman in South Carolina died from anaphylactic shock, a whole-body reaction, to being stung while she was gardening. In August 2008, a similar tragedy happened to a Seminole County Floridian, who stepped into an ant mound while walking his dog. In July 2013, a noted Atlanta, Georgia, woman died from a fire ant sting to which she was highly allergic. Every year, more than 40 Americans succumb to insect stings, such as those of fire ants, and 500,000 seek emergency room treatment.[3]

Origin, Introduction, and Spread

Native to parts of Peru, Brazil, and Argentina, the widespread invasive RIFA is a member of a large genus of stinging ants. It attacks humans, plants, and wild animals and disrupts environmental processes, particularly in the southern areas, where it gained a foothold in Mobile, Alabama, between 1933 and 1941. The aggressive RIFA joined another fire ant also from South America that had become established in the port of Mobile in 1918. This black imported fire ant (*S. rickteri*) was initially described as a subspecies of the RIFA but is now recognized as a separate species. A hybrid of the two species was recognized in the 1980s.[4]

Renowned Harvard entomologist and author Edward O. Wilson, who, as a high school student in his native Alabama, discovered and reported a new RIFA colony near Mobile's dockland, set about studying the new arrival. In 1949, the Alabama Department of Conservation hired Wilson, then a college undergraduate, and his friend to survey and assess the effects of the relatively unknown, nonnative insect on local agriculture. They speculated that the RIFA was already causing $178,000 worth of crop loss and damage in Mobile County and almost double that amount in nearby Baldwin County. The RIFA was clearly a marauder and needed to be destroyed, undergraduates concluded.[5]

Both the red and the black nonnative fire ants probably arrived in soil ballast on cargo ships. Once ashore, the RIFA, the more aggressive of the two, swept through Alabama and by 1953 had hitchhiked to Texas, where it infests moister eastern and central counties. Around this time the USDA completed a nationwide survey and noted that RIFA had already invaded 102 counties in 10 states. In an effort to stop its advance, federal quarantines restricted the shipment of plants, soil, and earth-moving equipment, together with straw, hay, and grasses in which ants hide, but these regulations failed to stop the insect's spread.

The RIFA ousts native ants and does this best in areas where average temperatures do not fall much below 10°F and where rainfall averages at least 10 inches per year. The species also likes irrigated fields. It has even pushed aside another nonnative ant, the Argentine ant (*Linepithema humile*), a versatile invasive insect that arrived in New Orleans sometime before 1891. The Argentine ant has also become widespread in the Southeast; however, contact with the RIFA has scattered its populations. This is just as well because researchers have discovered that the Argentine ant carries a virus that causes honeybee colony decline—a major issue worldwide with serious economic consequences.[6]

From Texas, the RIFA was carried farther west. A border inspection station in California intercepted the insects in 1984. Nevertheless, the RIFA marched on and took root in at least five counties in Southern California before heading onward, likely transported in soil or turf. Serious infestations broke out in Los Angeles, Riverside, and Orange Counties in 1998. The following year authorities launched a major campaign to get rid of RIFAs from 500,000 acres; it met with local but not overwhelming success. Fire ants turned up in California's Central Valley, and today state and county agents advise residents how to identify this invasive species and manage outbreaks. They publicize a list of 10 insecticide baits available for fire ant control.

The RIFA is already reported sporadically in Kentucky, Missouri, and Maryland, and experts predict a further push in California, New Mexico, Arizona, and Nevada, perhaps as far as Oregon, Washington, and Utah. The same is true in Texas, where the RIFA thrives in the eastern two-thirds of the state but has inhabited some sites in western areas. The RIFA extends to the average frost line in Oklahoma (and is likely to go 100 miles or so farther north as conditions change) and has also swarmed south into Mexico. The RIFA has spread from its original hold in Alabama as far as Maryland, where authorities remain on high alert. It is unevenly distributed along the Atlantic Coast, being likely to move north in Virginia and possibly into Delaware, although the black fire ant cousin appears better able to withstand colder temperatures. However, experts consider the black imported fire ant to be less of a problem because it occupies a much smaller area (northeastern Mississippi, northwestern Alabama, and southern Tennessee) and appears less aggressive and less mobile than its South American cousin.

Condemned as one of the world's 100 worst invasive species, the RIFA has gained entry to moist, warm lowlands in the Caribbean and beyond the New World. It has reached Taiwan, southern China, Hong Kong, the Philippines, and Australia.

Novel Adaptations: Hybrids and Multiple Queens

In Texas, the invasive RIFA lives in greater densities than those observed in South America because, unlike native fire ants, the two foreign species entered the United States and spread into Texas and elsewhere in the absence of natural predators. The invasive red and black species have formed hybrids, which are now well established in Mississippi, Alabama, and western Georgia. Recalling the saltcedar discussed earlier, we are reminded that interspecific hybrids are not limited to plants, and these crosses foster genetic combinations that may prove advantageous.

Experts have been trying to anticipate where the RIFA will end up, noting it has adapted to colder and drier areas than in its native range. An unexpected northward spread in the United States may be due to a lack of predators, as some suggest. Geographical similarity to conditions here with those in South America helped the RIFA consolidate its hold. It then went north and west, perhaps assisted by hybridization and also by changes in social structure, such as reduced territorial aggression because of a new behavior noted in the early 1970s of establishing multiple-queen colonies.

Multiple-queen, or polygyne, colonies bolster population growth with nests close to one another. Worker ants in the multiple-queen colonies are not aggressive about territory. According to University of Maryland biologist Matthew Fitzpatrick and colleagues, they "may allow fire ants to invade harsher environments by reducing both biotic resistance and extinction risk through increased abundance." However, pathogens can sweep through the densely packed nest mounds.[7]

The RIFA spreads when winged males and females (to be new queens) rise from a colony. Mating on the wing, the queens often settle within a mile or so of their starting point, often in bare, cleared patches. But pushed by favorable winds, they may fly much farther. After mating, the males die and the mated queens dig out depressions, lay fertilized eggs, and begin to establish colonies, whose mounds, essentially acting as microclimate incubators, take several months to develop. Colonies consist of a single queen, who may compete with other nearby queens and thereby limit the overall number of mounds to fewer than 300 per acre. Or she may be one among many queens using the same colony. In that instance, the mounds that contain many queens spread out more densely. Unfortunately, polygyne-type ant colonies dominate the RIFA range in Texas, and treating them can run to $45 per acre using livestock-friendly baits.

Controls: Baits and Insecticides

Charles Barr, fire ant specialist with the Texas A&M University System, has tested the effectiveness of a range of baits and insecticides on ant mounds. Baits are designed to interrupt nerve transmissions, body growth, and the metabolism of food. Insecticides act on contact with ants, altering respiration, nerve cell membranes, and nerve activities.[8]

Pesticides are used most widely to kill ant queens in their mounds. Drenching underground cavities with poisons or broadcasting chemicals, such as synthetic pyrethrum compounds, and setting out baits that worker ants carry belowground are approved methods used to eradicate the RIFA. However, they are expensive, can take weeks or months to work, and often provide temporary relief. Many insecticides also kill other animals, including beneficial insects, such as native ants. More specific baits such as Amdro are designed for workers to carry underground and feed to the queen. This delayed toxicity is most effective in single queen colonies, whereas in multiple ones, some queens probably survive.

Control: GPS and Sniffer Dogs in Australia

To understand the latest technique in locating and chemically treating the RIFA, we explore a potentially devastating outbreak in Australia that appears to have been of US origin. The RIFA turned up in two sites near Brisbane in February 2001. Queensland authorities acted promptly to eradicate this invasive species believed to have arrived in shipping containers. They also discovered a multiple-queen colony form of the invasive insect in Brisbane's southwestern suburbs, considering its presence in warm, moist northeastern Australia to be a serious matter.

Revising their estimate that the ants had been around for some time, officials set up a National Red Imported Fire Ant Eradication Program. With help from US experts, Queensland authorities sprayed an area in which there were 70,000 houses four times a year for three consecutive years. About A$200 million later, state personnel concluded that they had beaten back but not entirely eradicated the pests. UT expert Rob Plowes checked out the Brisbane fire ant problem and commended authorities for acting so energetically. They hit a new flare-up in 2013. But evidently the RIFA still exists in Australia and is on the move.[9]

Currently, aerial monitoring and GPS positioning pinpoint possible ant colonies. Then specially trained dogs sniff out live RIFAs. People inject the ant mounds with pesticides and bait them with poisons. However, in spite of such intense efforts, a RIFA outbreak appeared in Botany Bay, near Sydney, 450 miles south of Brisbane. Inspectors randomly checked out the lawns and backyards of 600 homes within a one-mile radius of the breakout site in December 2014. It was not clear how the ants arrived in Port Botany, most likely aboard ships. It was the first report from New South Wales.[10]

Phorid Flies, Pathogens, and Viruses

At the Brack Tract, researchers are pursuing an entirely different technique to slow the spread of RIFA. Larry Gilbert and fellow research scientist Rob Plowes, codirector of the invasive species program, have spearheaded interdisciplinary research on natural predators of the fire ant as far back as 1981, when RIFA first invaded the field lab. Rather than kill ants with chemical pesticides, they seek predatory insects to attack them, as well as fungi and viruses that will weaken and destroy the colonies. Their goal is to pressure the nonnative fire ants with as many biological weapons as possible. This will give native ants (including native fire ants) and many animals that fire ants prey on, or compete with, the chance to reestablish populations.

Finding an ant predator is not as easy as it seems. Water hyacinth, hydrilla, and saltcedar are large organisms that are easy to propagate and study, whether in the lab or field. As individual plants, they more or less stay put where they grow and within a given population of the same species are more or less undifferentiated from each other. Even so, finding successful biocontrol agents for them took time and much research. Ants, of course, are animals. Tiny insects, they can scurry away and hide in their labyrinthine burrows. Living in societies of a quarter million inhabitants, they are famous for their complex social orders, castes, and role divisions.

Not surprisingly, experiments to discover diseases and natural predators as biological or natural controls for fire ants have been intricate, complicated, and prolonged. They are continuing. Researchers have inspected potential parasites, pathogens, diseases, competitors, and predators—several of which show promise. For example, approximately 20 species of tiny flies in the genus *Pseudacteon* parasitize RIFAs in South America. This genus is one of about 230 phorid flies within the family Phoridae. Phorid flies that are being released to attack RIFAs in the United States are raised in the Brackenridge Field Station in Austin and until recently at USDA labs in Gainesville, Florida. The program is having success.

The first of these ant-killing flies, each one only one-sixteenth-inch long, were set free over RIFA colonies in Texas (in Travis, Brazos, and Dallas Counties), then at more and more sites across the US South, including the source area of RIFAs in Mobile, where three predatory fly species are now established.[11]

With a lightning strike, the female phorid fly, known as an ant-decapitating or brain-eating fly, deposits her egg into the thorax of a single RIFA. Her attack stuns but does not kill the ant, which scurries to its underground colony. Upon hatching, however, the fly larva feeds inside the ant and migrates to the host's head, devouring the head contents, until eventually the head falls off what is now a zombie ant host. The larva pupates in the empty head and emerges as a fly after about two weeks. The cycle from first attack to hatching and flight takes 30–45 days.

Recently an abundant, well-distributed, and highly parasitic phorid fly (*P. nocens*) from Argentina has shown promise. At two sites researchers from

(left) A phorid fly attacking a fire ant. (right) An adult phorid fly emerges from an ant head after devouring its contents as a larva. Both photos by Sanford D. Porter, US Department of Agriculture, Agricultural Research Service.

the University of Texas at Austin set up a novel release method. They buried imported *P. nocens* pupae in boxes from which the adult flies emerged and began to look for RIFAs. This attack species is most active in the evening and early morning when no other flies are about, which makes it attractive to use.

The aim is to release the predatory flies with differing behaviors and time cycles so that some attack smaller worker ants, others larger ones; some hunt in the morning or, as mentioned, others in the evening. So far, experiments have shown that these attack flies are host specific, meaning they go after RIFAs and not native ant species. Neither do the eight or so temperate-zone phorid predators of native US fire ants switch to alien invasive RIFAs. Researchers are confident that the six South American species released so far are aiming at RIFAs in Texas and in the additional southern states where the invasive species abound.[12]

The ant-eating phorid flies hover over their prey before they strike. Although the flying insects parasitize only between 1 and 3 percent of individuals in a colony, their buzzing causes major panic on the ground. Worker ants recognize the aerial predators and rush underground or seek to hide somewhere before emerging to forage. The flies disrupt colony behavior; as a consequence, native ants and other insects stand to benefit by having greater access to food and thus are better able to cope with the foreign ants.

Another control mechanism in the form of a spore-producing pathogen, *Kneallhazia solenopsae*, infects more than 10 percent of the imported fire ants in Argentina. It was first identified in Florida in the mid-1990s and has since spread into RIFA colonies in the United States, both naturally and with human help. Efforts involve attaching the spores to phorid flies to increase the likelihood of worker ants carrying the pathogen into their mound. Researchers are seeking other vectors for the spores. *Kneallhazia* shortens ants' lives and causes the multiple-colony queens to stop laying eggs, thereby lowering the number of ants in a colony until it dies.

Research on additional ways to manage and control fire ants is ongoing. Three viruses are known to infect RIFAs. Lab tests show that one of them causes high mortality, is easily produced, can be used on baits, and, importantly, is host specific. Researchers have also looked at wasps, mites, and nematodes as possible biological weapons against RIFAs. Contraceptive and toxicant baits tried out in a fire ant outbreak in Australia seem to have been effective. The old-fashioned use of boiling water poured into a colony can kill a colony provided it scalds the queen.[13]

RIFA Interactions with Native Texas Animals

People speculate about the connections among RIFAs and other animals. For example, discussions about the decline of the horned lizard, or "horny toad," throughout much of Texas have focused on whether RIFAs harass and kill young horned lizards as well as the lizard's chief prey, native harvester ants. It seems likely that insecticides used against RIFAs may have hastened the demise of these iconic lizards. Some people note, however, that the native horned lizard disappeared from places not inhabited by the invasive ants. Therefore, they remain unsure whether a single cause, such as the RIFA, explains the diminution of these once abundant and popular reptiles.

Some wonder whether the armadillo has a significant impact on red fire ant colonies. The animals' ranges overlap in Brazil, and observers have reported nine-banded armadillos, the most northerly existing member of the family that entered in Texas from Mexico in the late 1800s, tearing into ant mounds. Whether these insectivores feed often enough on fire ants, notably on their eggs and larvae, and get to a queen to control the RIFA population size is doubtful.

Except in Texas and Florida, armadillo populations remain relatively localized in the US South, especially in the Southeast where fire ants thrive. Research has yet to reveal whether fire ant densities are lower in areas where armadillos forage. However, recent research that appeared in a journal devoted to the welfare of the purple martin documents that breeding martins swoop into swarms of imported fire ants and grab mouthfuls of the flying ants and feed them to their young. Numbers of these acrobatic swallows are highest in the US South, notably from East Texas through states adjacent to the Mississippi River. This is also a key area where RIFAs swarm. Could it be that birds, such as swallows and swifts, more than mammals or reptiles, are having an impact on these invasive ants?

Ants as Drivers of Worldwide Environmental Change

One thing on which experts do agree is that we may be able to control RIFAs in varying degrees but never eradicate them in Texas or the nation. The invasive insects are with us to stay. Walter Tschinkel, a fire ant expert, likens the imported fire ant as the animal equivalent of a dandelion, doing best in the places humans disturb. However, while RIFAs are well-known occupants of disturbed soils and vegetation, the invasive insects do not do as well in disturbed places after they stabilize and recover. Tschinkel and coworker Joshua King suggest that if ecological recovery is allowed to proceed, then the invasive ant colonies are likely to be reduced in number and hopefully disappear.[14]

UT's Rob Plowes agrees but goes a step further. He insists that RIFAs are more than passengers in a car; they can actually drive it. He notes that the ants are associated with disturbance but can also invade undisturbed areas and cause biotic changes there. That is the meaning of fire ants being drivers of change. RIFAs feed intensively on a range of plants, tunneling into roots, stems, and seeds; and they aggressively attack small animals, such as other ants, spiders, lizards, frogs, mammals, including young livestock, and birds. In Texas, webcams have recorded RIFAs swarming into the nests of an endangered warbler and a vireo and killing their chicks.[15]

Characterizing the RIFA as the existing 300-pound ant gorilla in Texas ecosystems, UT researcher Ed LeBrun basically agrees with Plowes. By making ecosystems less diverse and less productive over time, the RIFA is driving environmental change. It infests croplands, pastures, parks, golf courses, home gardens, and cleared lots—on a county and regional basis.

"Worse" Newcomers in Texas

Foreign ants are not done with Texas. More have been added to the state list, including the Argentine ant (first discovered in New Orleans, Louisiana, in 1891 and now living in eight additional states, including Texas) and a newer pest called the tawny, or Rasberry, crazy ant, also from South America, which turned up in 2002.

Tawny Crazy Ant (*Nylanderia fulva*) - pupae and worker
Laboratory colony at The University of Texas at Austin, Brackenridge Field Laboratory.

Public domain image by Alex Wild and Ed LeBrun produced as part of the Insects Unlocked project at The University of Texas at Austin

The tawny crazy or Rasberry crazy ant. Photo by Alex Wild and Ed LeBrun, University of Texas at Austin's Insects Unlocked project (CC0 1.0).

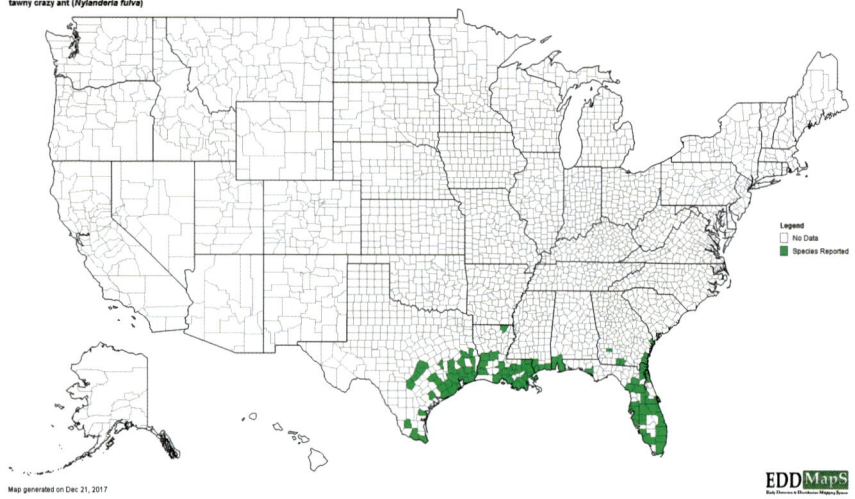

Legend
☐ No Data
■ Species Reported

Map generated on Dec 21, 2017

US distribution of tawny crazy ant (*Nylanderia fulva*). EDDMapS, University of Georgia—Center for Invasive Species and Ecosystem Health, 2017, http://www.eddmaps.org/.

The Crazy Ant

The tawny crazy ant (*Nylanderia fulva*, identified as such in 2012) is the most recent and even bigger ant "gorilla" in the Texas landscape. Edward LeBrun, who sees it that way, declares this ant to be worse than the RIFA. The crazy ant—also called the Rasberry crazy ant, after exterminator Tom Rasberry, who discovered it in Houston; and also the tawny crazy ant, after its color—hails from the same area as the RIFA, southern Brazil and northern Argentina. Named "crazy" for its erratic and frenzied-looking movements, the one-eighth-inch-long red-brown ant has recently become invasive in Texas and is also busily entrenching itself in three additional states, notably Florida.

Invasiveness comes not from bites or stings but from the crazy ant's love for swarming into new sites, even electrical equipment; sucking on plant nectar in various habitats; and filling the eyes, nostrils, and mouths of livestock to the point of suffocation. Just like the RIFA, the newcomer displaces native wildlife, disturbs nesting birds, and overwhelms native ants, including the stinging RIFA. Ant specialist LeBrun suggests that this recently arrived ant, smaller than the RIFA, exploits resources more completely than other ants.

It is displacing the RIFA, LeBrun says, by waging "chemical warfare" that enables it to detoxify the venom of RIFAs. Crazy ants have been observed taking command of RIFA mounds with some tenants still inside. They do this by secreting a formic acid shield that protects each one from fire ant venom. Consequently, when the two ants meet, the crazy ant replaces the RIFA and is now a top-order Texas pest, building colossal numbers in colonies that appear to run together. Neither the RIFA nor any other ant species can survive in close contact with crazy ant omnivores unless external controls regulate them, as they do in Argentina, for example, where native ants compete with this species.[16]

Unfortunately, this pernicious new ant builds huge numbers, so that multiple-queen colonies can reach over hundreds or even thousands of acres. Though its rate of spread is relatively slow on the ground (it does not make nuptial flights), we extend a helping hand by transporting the insect in compost, potted plants, hay bales, and so on. So far, the crazy ant, about which much still needs to be learned, has already infested 27 counties in Texas and is on the move. The crazy ant is turning up in both urban areas and rural places, including disturbed grasslands and wooded areas, but also forested floodplains. RIFAs are not common in these heavily forested areas, but crazy ants appear to fit right in.

Experts like Rob Plowes know much more about RIFA biology, about their natural predators and how to raise and release them. They know RIFA workers will carry poison baits into their mounds and, although aggressive and venomous, generally exist in lower densities than supercolony crazy ants. The crazies reduce the abundance of other ants by feeding on their larvae and brood stock. So, like RIFAs, the crazy ants reduce the diversity of other insects in any area. But numbers of them explode in summer and fall (rather than in spring for the RIFAs). Such astounding populations feed on the prey of other insect-eating wildlife, including mammals and birds, during movement and migration. It is the overall mass of the ants rather than the bites and venom they inflict that is most troublesome. Reportedly they emit a pheromone when stressed or injured, which brings massed cohorts to the site. Numbers increase exponentially as queens stay in place and produce more and more ants in adjacent colonies until they form a supercolony. They are like one big happy family, notes *Austin American-Statesman* staff writer Marty Toohey, after talking with ant man Edward LeBrun.[17] There is a sliver of hope, however. Crazy ant populations are dropping in Florida, perhaps due to natural pathogens that may be catching up with them, notes Plowes.[18]

Currently dogs are being trained in Texas to sniff out crazy ants, which are tinier, thus harder to detect than RIFAs, and do not make the usual mounds. This pinpointing provides possibilities for precise treatment. Chemicals are one way of dealing with them but tend to be expensive and short-lived. Another is to use phorid flies on them in a similar way as those that disrupt RIFA foraging patterns. Still yet another is to cultivate and release a fungus that infects crazy ant colonies. Tests are under way to determine whether the fungus will cause problems for other species.[19]

Little Fire Ant

Another so called tramp ant is the little fire ant (*Wasmannia auropunctata*), also known as the electric ant. It is one of a half dozen or so species that have been proficient hitchhikers and have established themselves and spread readily in new habitats worldwide. The little fire ant naturally occurs in both South and Central America and, as Rob Plowes notes, is a cryptic species, a group that has individuals that look identical to one another, and has spread as far north as Mexico and the Caribbean islands. It has been reported from Cameron County, Texas; Los Angeles County, California; 20 or more counties in Florida; and several Canadian provinces. This harmful ant has also been found in Australia, West Africa, and Europe (from where a native European fire ant, *Myrmica rubra*, has also made its way across to the US Northeast) and is fast becoming a menace on the Galápagos Islands, Hawaii (noted in 1999), and French Polynesia. First discovered in Tahiti in 2004, the aggressive little fire ant threatens to spread to 128 other French islands in the Pacific Ocean. Like its fellow RIFA, this small ant, which is barely one-sixteenth-inch long, packs a nasty sting and is a dangerous threat to any and all tropical island ecosystems. But unlike the RIFA, which prefers open, exposed sites and is not yet a resident in that region, the little fire ant has taken over moist shaded places, notably forest sites on the ground, and in trees, and lodges in walls. The opportunistic nester and food generalist is doing a lot of damage. This version of the RIFA, which is at least twice its size, is now on the list of the world's worst invasive species. Again, dogs are being trained, notes Rob Plowes, to identify and track down this tiny vicious insect, which has a painful sting, similar to that of the RIFA, once it has punctured the skin.[20]

The arrival and spread of nonnative ants, such as the RIFA, crazy ant, little fire ant, and other invasive tramp ants are easily overlooked. Unlike most of the species in this book, we did not introduce them on purpose. Adaptable insects hitch rides unnoticed and, once identified in a new site, need to be tackled immediately. This reminds us to be observant and to follow established procedures and practices to deal with them. Researchers are identifying modes of entry, key sites these ants prefer, and then stamping out outbreaks.

0.2 mm

The little fire ant or electric ant (*Wasmannia auropunctata*). Photo by Michael Branstetter, AntWeb.org (CC BY-SA 3.0).

The RIFA also shows that controlling an invasive species is both time consuming and expensive. Usual reliance on chemical pesticides neither eradicates this ant pest nor limits multiple, often serious, outbreaks. However, the technological sophistication of authorities in Australia demonstrates how tracking, pinpointing, and then drenching ant colonies with chemicals may effectively suppress this highly adaptable ant, provided populations are relatively localized and small. Concerns about the impact of these insecticides on nontarget species and the overall ecosystems always linger.

We continue to discuss the lasting effects of chemicals even in the context of second- or third-generation compounds. For example, recent discussion has increased about whether the most-used agricultural compound in the United States and likely the world, Roundup, a glyphosate weed killer pioneered by Monsanto, is being overly used, thereby building resistance in crop weeds. Debate has resurfaced about whether this product's "inert" ingredients may possibly be carcinogenic in humans.[21] These kinds of discussions highlight the effects on nontarget species, as well as lasting effects of synthetic pesticides. It has always been a trade-off between known and unknown side effects and the ease and convenience of using the chemicals.

Biological controls developed for RIFA management minimize such side effects in humans and the environment. In this instance, they provide us with invaluable insights about how to introduce natural predators from one organism's native range to areas where it is invasive and how to deploy such agents effectively.

University of Texas researchers have pointed out how host specific some predators, such as phorid flies, can be. Brackenridge Laboratory scientists are bent on experiments and necessary fieldwork to suppress RIFA populations as much as possible. By targeting the RIFA, they have generated protocols for introducing, rearing, and releasing the ant's killer flies. Practical experience, including a nuanced understanding of the localities and conditions that make this phorid species more effective as a control agent, has generated critical information that researchers publish and pass on. The study, experimentation, and release of predatory species highlight how consultation works—in this case between North America and South America. Such exchanges of knowledge and expertise are vital when coming to grips with other invasive ants, for example, and show how communication and collaboration are taking place on a global scale.

8 Texotics
Ranching Exotic Wildlife

Game animals from all parts of the world are now at home in Texas. Landowners keep them because they are pleasing to look at and are invaluable sources of income. However, escaping into the wild, these nonnative mammals add pressure on rangelands and native wildlife. Biologists are concerned about how some exotics may outcompete native deer, especially during drought. And as the population of nonnative deer, antelope, goat, and sheep has increased markedly in Texas over the past 50 years or so, some species are invasive. Like other plants and animals that we have introduced, raised, and protected, once individuals escape from confinement and begin to breed in the wild, control and management of them grow problematic.

Deer researchers, breeders, raisers, catchers, sellers, and hunters play important roles in Nina Leigh Vizcarrando's film *A Kind of Wild*. Shot for her master's of fine arts degree at the University of Texas at Austin in 2014, Vizcarrando offers a cross section of contemporary ranching in Texas, not with usual livestock but with exotic game animals. She shows how keeping so-called Texotics, that is, introduced nonnative game species, mostly mammals, is commercially important for many landowners, who provide trophy specimens to fee-paying hunters. She also shows how owners and managers manifest kindness toward their stock and treat individual animals with compassion. Therein lies the "paradox between hunting and conservation," she says, "as individuals and as a capitalist society our instincts to love and nurture are often intertwined with our desires to profit and own."[1]

Almost all of Texas is under private ownership. For generations, in order to make a living, property owners have farmed and ranched. In recent decades, a push to use the land more profitably, notably in the livestock-rich drier central-west counties, has motivated ranchers to run game animals with cattle, goats, and sheep. Increasing numbers of landowners have also specialized in raising nonnative game animals to bolster incomes from hunting.

Today, according to Charly Seale, rancher and executive director of the Exotic Wildlife Association, who represents 5,000 exotic game ranchers in North America, there are about 125 different species of exotic animals in

Texas. Most are game mammals from Africa and Asia and number several hundred thousand. They exist on hundreds of ranches. Landowners and managers offer packages to hunt them. Paying guests may shoot selected animals as trophies for the den wall and meat for food lockers. Owners also auction off surplus animals or trade them to other ranchers, who are looking for something new or different for their spreads.[2]

In theory, these animals are similar to livestock when safely fenced, fed, and checked for diseases. However, there are two important differences. Unlike cattle, horses, pigs, sheep, and goats, Texotics are in most cases not domesticated animals. They are wild animals. And when they escape and multiply, they have the potential to become invasive. Farmyard pigs *have turned into* feral hogs. Texotics are mostly wild to begin with and avoid contact with people. Second, many, if not most, Texotics are raised to be hunted. While the animals are classified as livestock, owners intended them to be treated as a source of trophies and meat. Given their extraordinary increase in numbers, continued spread across central and southern counties, the number of species now acclimated to the Lone Star State, and the fact that a few are endangered or extinct in their homelands and even in the wild, their presence raises fascinating questions.

From Deer Hunting to Exotic Game

Texans have always liked to hunt. Settlers trapped and shot black bear, deer, wild turkeys, and smaller animals to make ends meet. Old-timers supplied and bartered wild meat, bonding with colleagues and neighbors for the conviviality of the chase. Hunting was a way of getting to know places, becoming familiar with the land and its plants and wildlife. Boys learned field craft from relatives; their elders honed outdoor skills and trained dogs to chase and retrieve game. The hunt combined utility with pleasure. It still does.

Hunting game mammals, primarily white-tailed deer, the state's official large mammal, is a hallowed tradition. Every year from October 1 through early November (for archery) and from early November through mid-January (for firearms), hunters set out to bag more than 500,000 white-tailed deer out of an estimated population of 4 million animals. Texas ranks number one among states in selling more than 1.2 million deer licenses and is also number one in the number of animals harvested. Hunting generates about $1.5 billion annually, and this dollar amount climbs to more than $2 billion in Texas when rearing and selling deer and safari-type visitation and viewing on ranches are included. Most deer abundance comes from recruitment in the wild and recently from an increase in husbandry, that is, raising wild deer in the manner of livestock. Currently, about 110,000 whitetails are reared in Texas every year. Many are used to stock ranchlands or released where they have been bred during the hunting season for clients to shoot.

This growing but controversial practice of raising native deer, rather than relying on wild-bred ones, mirrors a tradition that began with nonnative animals in the mid-1920s. Reportedly in 1924, several nilgai antelope, native to central and North India, were introduced onto the 825,000-acre King Ranch in South Texas, one of the world's largest ranches. It spreads into parts of six counties between Corpus Christi and Brownsville. King Ranch foreman and wildlife specialist Caesar Kleberg took an interest in adding the nonnative nilgai acquired from the San Diego Zoo to the ranch's livestock operations. Kleberg pioneered what has become a practice of introducing and building up populations of foreign mammals in Texas; initially many came from zoos. Today, the King Ranch manages 10,000 nilgai antelope on its Norias Division and offers clients' day hunts at $850 per gun plus a harvest fee.[3]

In the 1930s several landowners added more nonnative mammals. In 1932, the fallow deer, native to the eastern Mediterranean Sea region, was released on the Blackjack Peninsula, later named the Aransas National Wildlife Refuge, a federal reserve dedicated to saving the remnant population of wintering whooping cranes in North America. During that decade, Charles Schreiner III, owner and operator of the famous Y-O Ranch, itself an important source of nonnative animals, noted that Richard Friedrich was

Nilgai antelope were one of the first large exotic game animals to be introduced to Texas in the 1920s. Photo by Hatem Moushir (CC BY-SA 3.0).

a driving force in introducing exotics. Friedrich fenced off a portion of his Bear Creek Ranch in Kerr County on the Edwards Plateau and stocked it with a dozen different species of deer and antelope he acquired with assistance from the San Antonio Zoo. This spread became the first exotic game ranch in Texas, if not the United States. Friedrich also passed on surplus stock to other ranchers. These animals included the axis, sambar, sika, and barasingha (swamp deer), all members of the deer family Cervidae, found in South Asia and East Asia.[4]

The rough limestone country in Central Texas grew into a major hearth in the United States for raising deer and other foreign mammals. David Rickenbacker put the Bear Creek Ranch on a paying basis after his father purchased it from Friedrich in 1951. Schreiner credits David for starting the practice of generating ranch income from paying guests wanting to hunt exotics. Schreiner recalls that his own Y-O Ranch in Mountain Home generated income from leasing land for hunting native deer and turkey as early as 1939. As income from cattle, goats, and sheep declined during the 1950s drought, the Y-O turned to exotics for extra revenue, as clients were able to harvest these foreign species all year. Other ranchers followed suit. It was, Schreiner recalled, a year-round option for hunters wanting something new.

The graceful blackbuck antelope, which belongs to the cattle, antelope, and sheep family Bovidae, like the nilgai, was another early successful introduction. Both cloven-hoofed ruminants have done especially well in South and Central Texas, although prolonged freezes or snow on the ground may cause die-offs, especially where they depend on natural forage. There are now some 34 different antelope species, subspecies, and breeds in Texas and 15 or so similar sets of sheep and goats. In recent decades, interested owners have lengthened this list to include members of the zebra and horse family, Equidae. Additional newcomers, such as giraffes, rhinos, llamas, and camels, are also regularly featured in drive-through safari parks.

Increasing Popularity and Growing Concern

Coordinated and collaborative interest in both the economic and environmental costs and benefits surrounding the introduction of exotic game animals dates back 50 years. In 1967, the Texas Chapter of the Wildlife Society teamed up with the American Institute for Biological Sciences to hold a symposium in College Station, titled "Introduction of Exotic Animals." Federal, state, university, and landowner participants debated the recreational, commercial, and environmental merits of owning Texotics and recognized that populations were growing.

The keynote speaker at the 1967 symposium was Gardiner Bump, biologist in charge of the State-Federal Cooperative Foreign Game Program, established in 1948. The US Fish and Wildlife Service organized this program to introduce new game animals into habitats that lacked healthy populations of native species. Biologists investigated the possibilities for many foreign species and eventually released about 20 potentially useful ones, mostly in the US South and Hawaii. The program was discontinued in 1970. In justifying the Foreign Game Program, Bump criticized a traditional hit-and-miss approach in releasing alien mammals and birds. He claimed that many introductions had failed to take, including in Texas, because of the poor quality of stock, a dearth of organization or oversight, and misunderstandings about the animals' habits and requirements. Bump was talking about game birds, though what he said also applied equally to foreign mammals (no mammal biologist worked in the Foreign Game Program at that time).

Based on the premise that human activities have transformed or obliterated the habitats of many native species, "filling vacant or drastically understocked habitats was a worthy goal" Bump argued. After comparing similarities of climate and terrain in several US states, 26 species of foreign game birds, such as pheasants, francolins, and tinamous (out of a list of 150 or so), had been recommended for field trials. The same climatic and terrain variables have served to help identify suitable ranges for African and Asian animals in Texas.[5]

The Foreign Game program, like the USDA's Bureau of Plant Industry, which promoted the cultivation of Chinese tallow throughout the South, was committed to introducing individual species, not determining whether the natural landscape, or what remained of it, could be better restocked with native ones. The goal was more about maximizing yield, economic viability, and sport than about ecosystem integrity, which at that time was still a novel concept. A similar viewpoint had drawn together a group of landowners, who founded the Exotic Wildlife Association also in 1967. The association, based in Ingram, Kerr County, is committed "to encourage and expand the conservation of both native and non-native hoofstock, and to help members develop and strengthen markets for their animals." The association's 3,700 members raise and sell scores of introduced animals while reinforcing the right to hunt them on private ranches. A state legislative ruling makes these nonnative species essentially domestic livestock; landowners can acquire, release, and hunt exotics on their own properties at any time of the year.[6]

Aware that numbers of exotic animals were booming and that more and more ranches were stocking them, Texas Parks and Wildlife biologists began periodic censuses. Personnel had become concerned about how the nonnative mammals were interacting with native wildlife. Biologists were especially interested in animals that had escaped from ranches and were reproducing in the wild. The first survey in 1963 recorded 13 species totaling 13,000 animals on 178 ranches. A total of 142, or 80 percent, of the ranches had started with exotics after 1950 because of falling prices for livestock during the severe drought of that decade. By 1971, the number had virtually tripled to 35 species consisting of 45,700 animals. Ten years later, 39 species and about 57,000 animals existed on 316 ranches spanning 4.5 million acres, mostly in Central and South Texas. Axis deer, mouflon-Barbados sheep, blackbuck antelope, fallow deer, auodad sheep, sika deer, and nilgai antelope, in that order, dominated the tally.[7]

By 1990, the population of Texotics had increased almost threefold to 164,000 animals, and a large number had escaped and were free ranging. At that time, Texas Parks and Wildlife personnel had information concerning 67 species on 486 ranches totaling 2.3 million acres in 137 counties, mostly on the Edwards Plateau and in South Texas. Clearly, the nonnative animals were thriving; Texas was starting to resemble a combination of Africa and Asia.[8]

Texotic numbers have continued to increase. In a 2007 *Texas Parks and Wildlife* magazine article, Rusty Middleton noted that state biologists speculated that a minimum of 275,000 and a maximum of 1 million nonnative animals belonging to at least 76 species lived in the Lone Star State. Most of them were in Central Texas. Some 3,700 operations in Texas generated upward of $1 billion in economic activity and more than 14,000 jobs. No other state can boast these kinds or numbers, Middleton reported, nor the commitment to raising them. There are more exotic species in larger populations in the Lone Star State than anywhere else, including California and New Mexico—two additional states with traditions of holding sizable populations of nonnative mammals.[9]

Competition with Native Wildlife

Concern about the ecological risks as well as the economic benefits of mixing foreign mammals with native animals has resulted in Texas essentially becoming a huge outdoor laboratory for experimenting and observing how nonnative species interact with one another and with native species. In the late 1960s, researchers and ranchers wanted to know what happens when the nonnative animals associate and compete with the native white-tailed deer. And what happens to the land's capacity to sustain wildlife in general when free-ranging exotic species seek forage in drier-than-usual conditions?

Both queries address the issues of invasiveness in its environmental context.

The response to the first question about competition is that as often as not, the native deer lose out. There are several reasons for this. Introduced African and Asian ungulates are versatile feeders and are often able to adjust their intake to what circumstances and season allow. Historically, Central Texas rangeland, where the majority of Texotics are located, has been overgrazed. Competition with existing livestock and native deer for limited food is an issue that the foreign animals have to deal with.

Experiments suggest exotics appear to cope well, too well in fact. For example, six whitetails and six sika deer were placed in a 96-acre enclosure, and six whitetails and six axis deer were placed in a similar one. They remained inside the large enclosures for nine years after which researchers checked the two sites. They discovered three whitetails and 62 sika deer in the first enclosure and three whitetails and 15 axis deer in the second enclosure. Both exotic deer species had flourished at the expense of the white-tailed deer without any human interference. The reason was that these exotic species, and additional ones, prefer forbs and browse but will switch to grass when they have to. White-tailed deer do not switch to grass and suffer malnutrition. Thus, native deer are less able to survive food shortages, caused by drought, for example, which occurred during the nine-year experiment. The exotic animals depleted the forage essential for the whitetails and then adjusted their feeding (by taking grasses) as conditions grew worse.[10]

When a landowner intends to raise both exotic and native deer, there is no one size fits all, as the saying goes. "It's all about keeping the habitat healthy and the population in check," notes state biologist Mitch Lockwood. Keeping both native and nonnative species is tricky and gets trickier when cattle and sheep and goats are in the mix. Deer and livestock have different habitat preferences, food choices, and behaviors, Lockwood notes. Therefore, each landowner needs to figure out what the land can sustain. Lockwood notes that Texas Parks and Wildlife personnel are ready to assist with calculating the carrying capacities for various game species in a specific landscape, whether mixed in or not with livestock or native deer.[11]

The movement and dispersal of "escapees," individual exotic animals that jump over, get through by means of a hole opened by feral hogs, or scramble under a fence due to a freshet, intensifies competition. Escaped individuals interact with their own kind and other exotic animals as well as native deer and local livestock. For example, the axis deer, originally from Sri Lanka and peninsular India, inhabits at least 92 counties in Texas, mostly on the Edwards Plateau. It is known as *the* Hill Country exotic and is most populous and among the most handsome deer. There are at least 40,000 in Texas, and several thousand are free ranging. Hunters prize the axis deer. It makes a well-regarded trophy and affords lean meat without the gamey flavor of whitetail venison on a carcass that may weigh 200 pounds. However, once free, axis deer often travel in herds, trespassing as it were from pasture to pasture and ranch to ranch. They are unwelcome, for example, on Selah near Johnson City, which David Bamberger has sought to restore to a natural state. Axis deer encroach onto the 5,500-acre ranch preserve and disrupt habitat-restoration efforts by their herding and numbers.

Many people regard axis deer as direct competitors with native deer and argue they are more resistant to disease, placing the native whitetails at a competitive disadvantage. Although axis deer are primarily grass eaters, and in theory compatible with whitetails, which prefer browse, with numbers increasing so that each animal nibbles just a little browse, these nonnative deer diminish the food of native deer. Hunters argue that extra pressure is needed to keep axis populations in check. Dietary overlap studies show that there are at least five exotics, including sika, fallow, and axis deer, that also compete for forage across the Edwards Plateau. This justifies targeting them to keep native deer healthy.

Competition also involves exotics beyond Central Texas. Aoudad or Barbary sheep, native to North Africa, were introduced by the State (at the

An adult male axis deer or chital. The axis deer is arguably the most populous exotic game animal in Texas. Photo by T. R. Shankar Raman (CC BY-SA 4.0).

Aoudad or Barbary sheep. Native to Africa, aoudads compete with native desert bighorns and mule deer. Photo by Rei (CC BY-SA 3.0).

behest of local landowners) into Palo Duro Canyon in 1957. Six years later, sheep numbers had grown to merit hunting. Adapted to arid conditions and rugged canyons, the population continued to increase. By 1990, the state population reportedly topped 20,000 aoudads, which had expanded into dry portions of the Edwards Plateau and the Trans-Pecos, where studies found them competing with native desert bighorn sheep and mule deer. As desert animals and opportunistic feeders, the aoudad has pressured the recovery of the bighorn sheep.

Exotic Diseases

The transmission and spread of disease are other concerns about exotics. The first case of malignant catarrhal fever (MCF) in North America was reported in axis deer in Texas in 1968. It is a cell-associated herpesvirus passed to deer, notably axis and whitetails, by domestic sheep. The buildup and continued uncontrolled spread of nonnative species poses considerable risks of individuals being either asymptomatic carriers of a disease or becoming infected from interacting with domestic sheep and goats.[12]

Elaeophorosis, a parasitic nematode transmitted by horse flies that causes lesions, tumor masses, horn deformities, instability, and eventual death in certain deer and livestock, appeared in sika deer in Texas in 1978. It was believed to have originated in whitetail hosts that had been identified as carrying it asymptomatically on occasions in the US Southeast, but not at that time in Texas. Chronic wasting disease, a disease of the nervous system that leads to weight loss, lethargy, neurological changes, and eventual death in deer, appeared in far West Texas in 2012. This transmissible disease among deer, elk, and moose has been reported in 23 states and two Canadian provinces. People are watching for signs of it among deer populations in Texas. So far there have been few outbreaks and none among exotic deer; however, biologists and land managers are keeping a careful watch for the possibility of it spreading in Texas. It has been known to infect red deer, and it can be experimentally transmitted to fallow deer.[13]

In 2007, Lynn Brezosky reported 37 nilgai antelope shot from a helicopter in the Lower Rio Grande Valley National Wildlife Refuge. The animals had drifted in from the north, and in addition to their trampling and competing for forage, federal officials feared an outbreak of tick fever from ticks the nilgai had picked up. They had no option, they said, but to hire a chopper and shooter to kill the foreign animals because there was no way to physically check out the exotic mammals along the border with Mexico. USDA inspection personnel have waged a battle and are on constant alert to keep two species of ticks proven deadly to cattle from crossing the border. However, in 2017, AgriLife experts reported an outbreak of tick fever in South Texas that resulted in 500,000 acres being placed under quarantine. Named after outbreaks resulting from Longhorn cattle drives in the 1880s, eradication programs cleared the area by the 1940s, establishing a buffer zone along the Rio Grande. Although cattle remain the major host, native deer and exotic ungulates are also hosts to ticks. Nilgai are a known host for the cattle tick in India, and thus entomologist Pete Teel noted, "So what we've done is bring both the ticks and nilgai together again." There is no cure for Texas tick fever.[14]

In another case, ticks known to be vectors of heartwater disease, a deadly African disease not found in the United States, were taken from three black rhinoceroses. The animals were confined in South Texas but had passed through a period of quarantine. The disease is deadly to cattle, deer, sheep, and goats.[15]

Hybrids: Pros and Cons

Hybridization also poses a dilemma. On the one hand, breeders offer attractive animal hybrids as trophies, especially between goats and sheep. The so-called Texas hybrid ibex, a cross between one or more wild goats, such as the Alpine, Nubian, Spanish, Siberian, or Walia, and a subspecies of domestic goat, is an attractive one. The Corsican sheep or ram is another. This trophy animal is a European mouflon sheep, native to Corsica and Sardinia, crossed with the Barbados sheep and sometimes other sheep, and was developed on the Y-O Ranch. It is a registered trademark of the Y-O and makes a very desirable trophy. Males are usually brown with long guard hairs and a mane but may be black (Hawaiian black sheep) or white (Texas dall sheep), and have fully curled horns that turn outward at the tip. Other special breeds are popular, such as the painted desert sheep. It is a cross between a European mouflon sheep and a Rambouillet sheep and has distinctive coloration and large horns.

Antelopes have also generated stunning hybrids. By breeding scimitar-horned oryx females with gemsbok males, ranchers have been producing the "scimbok"—another desirable trophy animal. Reportedly the scimbok is able to reproduce, as are hybrids produced by the scimitar-horned oryx with other oryx (Arabian, Beisa, and fringe-eared) and the addax. Beauty and uniqueness doubtless increase their allure.

However, with the explosion of hybrid saltcedar in mind, or the novel crosses between red and black invasive fire ants, what happens when these animals and their hybrids escape, go wild, and reproduce? Ranchers have come across nonnative animals that "don't quite look right." It happens with transcaspian urial sheep, which are high-strung and skittish but will interbreed with domestic and other sheep. The same is true for Iranian red sheep, a native hybrid population found in the Alborz Mountains in Iran. Although very wary, these red sheep are fertile when placed with other sheep. Many of these animals, as are ibex and other goats, are difficult to confine. When stressed, an individual may scale a 10- or 12-foot-high fence; they are agile climbers and conceal themselves in draws and rocky terrain.

Jests take place in web chat rooms about these and other kinds of hybrids being used to blur genetic lines to undermine or thwart regulations about hunting or possessing various species. New engineered breeds add to the mix of existing food needs and behaviors that exotic animal species express in Central Texas.[16]

Finally, knowledge and understanding of the life history and ecological roles played by nonnative species in Central Texas are still being generated. Several mammal species, such as the nilgai, blackbuck antelope, and axis deer, have been carefully studied. But many of the other nonnative species remain less well known. We have seen repeatedly that we cannot fully predict how exotic species, much less their novel crosses, will behave in new places.

Today, there are more white-tailed deer in the Hill Country than else-where in Texas, which is arguably comforting news. With a population density of 113 deer per 1,000 acres, four times the state average, the 2.1 million white-tailed deer must forage with thousands of free-ranging nonnative mammals and also comingle with the main bulk of exotic hoofed mammals on fenced ranches. This suggests they are doing well even with the Texotics, but environmental change can sully this picture. We recall how simply delaying the flooding regimes of rivers in the Southwest by a few months unknowingly allowed saltcedar to crowd out cottonwood and willow woodlands. We have seen that drought puts native deer at a competitive disadvantage among free-ranging alien deer.

Conservation or Exploitation?

Just as the discovery of the endangered southwestern willow flycatcher nesting in invasive saltcedar thickets made us modify plans about how to control the tree, so does the realization that some Texotics are endangered in their homelands add moral complexity to the exotic game story.

The scimitar-horned oryx, for instance, is extinct in the wild. This stunning antelope, with backward-curving horns that can reach four feet in length, was once widespread across North Africa. Around the year 2000, after years of habitat loss and overzealous hunting, it could no longer be found in the wild and existed only in zoos and captive-breeding programs. The recent reintroduction of a small herd into a reserve in Chad may raise its official status from extinct to critically endangered. If scimitar-horned oryx flourish on Texas soil—and 11,000 of the creatures are currently scattered across Texas ranches—this is arguably a step in the right direction.[17]

Recently, the Exotic Wildlife Association has been using its clout with authorities in South Africa to negotiate the importation of 1,000 white rhinos into South Texas. The aim is to protect these threatened animals, not hunt them. Demand for rhino horn and organized international criminal syndicates involved in rhino poaching have increased in recent years throughout the animal's native range. The association is well aware of the poaching and hopes that introducing the rhinos to similar habitats in Texas will help bolster numbers so that eventually some of them can be returned to their homeland.[18]

Such restocking has already taken place with Texas-raised blackbuck antelope, some of which have been reintroduced to their native range in

Scimitar-horned oryx. Once widespread in North Africa, this oryx is now extinct in the wild, though recent introductions from captive-breeding programs hold promise. Photo by FisherQueen (CC BY-SA 3.0).

Pakistan. Numbering in the millions, blackbuck once inhabited the whole Indian subcontinent south of the Himalayas. Their range decreased sharply in the twentieth century because of unsustainable hunting, and they are now extinct in Bangladesh and Pakistan. Introduced to Texas in 1932, blackbuck became the second most populous exotic animal in the state (after axis deer) by 1988. On two occasions, in 1970 and 1985, a group of Texas-raised blackbuck was returned to that subcontinent for reintroduction.[19]

Some dislike the hunting. The Friends of Animals, a nonprofit, international animal advocacy organization, wants to stop the hunting of endangered African animals, notably the scimitar-horned oryx, and aims to have them only in their native ranges. It is hard to disagree with their argument in theory. If the bald eagle, listed as a US endangered species from 1967 to 1995, were raised in European reserves for hunters to shoot at $1,000 a head, most of us would rankle at the practice. But if these same reserves produced thousands of eagles for repopulating the bird's native habitat, well, there's the rub. Many see hunting as the best way to conserve endangered animals and point to growing populations of the oryx and blackbuck in Texas as proof of their viability over the longer term. Without hunting,

Blackbuck antelope, adult stag. Blackbuck are threatened in their homeland on the Indian subcontinent. Photo by Chinmayisk (CC BY-SA 3.0).

they argue, there would not be the same economic impetus to raise these creatures.

Hunting Texotics is big business. For instance, a hunting package for axis deer (one of 13 packages for different exotic mammals) on a ranch a 2.5-hour drive from Austin consists of a three-day, two-night stay with complimentary meals, bar, and one's own field guide. The baseline price is $3,250–$3,800 for the deer. The axis deer is not threatened because currently there are no rangewide threats to the animal and many live in protected areas. But for rarer, more unusual, or simply larger creatures, that price increases: $4,000 for a scimitar-horned oryx, $6,000 for a gemsbok, and as much as $21,000 for a top-quality elk.

A glance at any website turns up more than 100 ranches offering similar exotic-game hunting packages in Central Texas, most with multiple species and individual animals well suited for the trophy hunter. Some ranches claim to keep or own 30 or more species of exotics, including zebra, wildebeest, and the rare "tres amigos"—dama gazelle, addax, and scimitar-horned oryx. They buy and sell stock to other ranches, zoos, and animal parks. Others include bird hunting and fishing. Almost all provide archery and crossbow hunting as well as the use of conventional firearms, including pistols.

Whether the number of exotics results in either the retention or improvement of habitat is inconclusively debated. The incentive of fee hunting motivates private landowners to invest in quality stock and conserve ranch habitats. However, such fees may also result in supplemental feeding as managers run more animals than the land can support naturally. Thus, raising and keeping exotics may degrade the environment, especially in harsh conditions. Whatever position you take, one thing is clear, Texotics are increasingly important, if not vital, sources of revenue for landowners and managers.

Hunting these nonnative species has become a way of life. As more and more exotics come in and are released, and information about raising them increases, the economic benefits of having so many foreign trophy-type animals grows. Elizabeth Mungall's *Exotic Animal Field Guide* shows how popular they have grown, not only in Texas but also increasingly throughout North America. She describes 80 hoofed exotics, most of which live in Texas. Her 2007 *Guide* illustrates expensive trophies that generate important income on which many ranch managers have come to depend. They are also novelties that photographers pursue while taking safari-style outings.[20]

Today, you can find specialists who offer detailed advice to investors and owners potentially interested in running exotics. They guide you about which animals to select, which habitat is optimal for them, and what are the best stocking rates. They counsel about fences; supplemental feed; how to capture, treat, and transport them; and when and where to market them—any kind of question about Texotics you can imagine. Growing up with exotics, many ranchers have sorted out the nuts and bolts of keeping these foreign species. They have advanced a long way from releasing a few beasts from Africa that they bought as surplus from zoos. That was two or more generations ago. Today, hundreds of ranches stock scores of species, and breeders are keen to raise trophy animals and establish new ones. They claim managing exotics is good for business and for conservation. Some landowners also insist they are good stewards who respect their wild charges and treat them with dignity. Ranching them is humane, is aesthetically satisfying, and presents a puzzle and a challenge they welcome.

However, it is unclear as to how many exotics exist or how many, according to the song, are making a home where the buffalo roam, that is, are dispersing far and wide in Texas. The growing numbers of exotics that have escaped confinement and are breeding is posing problems for landowners, who see the foreign game species as hybridizing in the wild, interfering with livestock, consuming too much forage, and potentially spreading disease either from or to domestic animals and native wildlife. It is the uncontrolled multiplication and spread that causes the most concern. The number of exotic species and their populations make state biologists, who are keenly aware that even small changes to an ecosystem can have lasting results, alert about their invasiveness, especially when research shows that many nonnative game mammals survive better than native animals do during drought. Exotic game is getting along very well in Texas both inside and outside the fences designed to contain them. This section of the New World is looking more and more like parts of the Old World, as graceful, elegant animals, together with iconic and endangered ones, are staking a claim in being new Texans.

9 The Beat Goes On
New Challenges

We are playing a game of catchup with invasive species. In this globalized and interconnected world, more and more nonnative organisms are being spread around, forcing experts and the interested public in Texas to identify, interdict, and draw up plans for dealing with what is likely to arrive from distant parts of the world and prove invasive.

One of the newest threats to our state is the Brazilian peppertree, which reminds us that exotic garden ornamentals, seemingly harmless for decades, can suddenly show invasive qualities and proliferate. Aquatic organisms continue to play leading roles. Another Brazilian, giant salvinia, the latest and arguably worst aquatic plant to hit our state, grows at a rate that astounds veteran biologists. The zebra mussel, recent scourge of the Great Lakes, has journeyed from the Black and Caspian Seas to infest Texas lakes with its tiny sharp shells. And off our coasts, the red lionfish from the western Pacific Ocean is emptying our reefs and inshore waters of native fish, with only spear fishermen and our appetites to keep them in check.

Seemingly innocuous human activities continue to encourage dispersion. Shipping wooden pallets and transporting firewood have brought the petite Asian beetle known as the emerald ash borer to our doorstep, along with the demise of a great forest tree. And spelunkers and cave tourists returning from Europe are the likely vector of a fungus-type disease called white-nose syndrome that is threatening our bat populations. How far these new threats play out remains to be seen, but more and more people are being vigilant. These new arrivals are challenging Texans' collaborative ingenuity and collective know-how for monitoring the arrival of such deadly invasive species and establishing plans about managing them. Each organism has a peculiar set of characteristics that require knowledge, commitment, and remembering similar situations, strategies, or programs adopted in the past.

Brazilian Peppertree

As far as is known, Brazilian peppertree (*Schinus terebinthifolius*) first grew in Texas in the Rio Grande Valley in the 1940s. Robert Runyon, an accomplished amateur botanist and mayor of Brownsville, collected a cultivated specimen

(left) Brazilian peppertree encroaching on the Texas coast near Port Aransas. (right) Close-up of fruit. Photos by Robin Doughty.

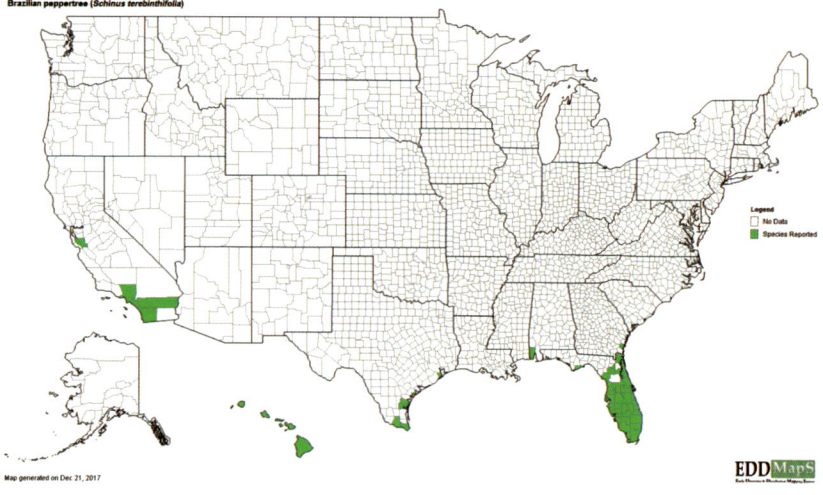

US distribution of Brazilian peppertree (*Schinus terebinthifolius*). EDDMapS, University of Georgia—Center for Invasive Species and Ecosystem Health, 2017, http://www.eddmaps.org/.

in that city in 1941. Two years later a professional botanist noted a 20-foot-high specimen growing in front of the Casa de Palmas Hotel in McAllen. By 1947 a peppertree was growing beside the noted resaca in the town of San Benito, 20 miles northwest of Brownsville, presumably self-sown.[1]

Peppertree plantings did not cause alarm. Gardeners have often relied on exotic ornamentals to add zing to a landscape. The showy evergreen perennial peppertree (reaching as high as 30 feet) bears stunning clusters of bright red berries, just in time for Christmas. Owners admired them. In fact, plant catalogs in New York touted the species, native to Brazil, Argentina, and Paraguay, as a garden ornamental as early as 1832, but it thrives only in warm climates. In Florida, for example, it grew so commonly in cityscapes in the first half of the twentieth century that people called it Florida holly. And in Hawaii, the seed clusters of this so-called Christmasberry decorated leis and hats. South Texas joined the celebration.[2]

Within a few decades, however, peppertrees showed signs of misadventure. In Florida, naturalized stands started appearing in the Keys in the 1950s and were no longer "well behaved," as gardeners say. The tree crowded into disturbed sites, such as highways, canals, fallow fields, and drained wetlands and even invaded slash-pine forests and woods on outer barrier islands. It spread into Everglades National Park by 1969 after birds fed on its seeds and transported them to remote hummocks. Although banned from commercial use in Florida in 1990, the tree grows in almost every plant habitat in the southern part of the state, covering at least 1,100 square miles.[3]

In Hawaii, and other places with the same type of climate, a similar pattern emerged. Peppertree marched up slopes on Oahu in the 1940s and acquired noxious-species status in the state in 1978. Experts consider the plant one of the most important nonnative species that threatens federally listed endangered native plants across Hawaii. The tree has also made serious inroads into the US Virgin Islands, Puerto Rico, and Southern California and has popped up in Alabama and Georgia. Outside the United States, Brazilian peppertree has naturalized in tropical and subtropical regions worldwide, including Bermuda, the Bahamas, the West Indies, Mediterranean Europe, southern Asia, and South Africa.

In Texas, biologists identified the tree growing wild around Corpus Christi in the 1990s and discovered it locally abundant on barrier islands and along the Laguna Madre. By 2000, it had crept into nearby Port Aransas and was settled on Mustang Island. Five years later state authorities added it to Texas' noxious-plant list, which makes it unlawful to sell or distribute the peppertrees in the state. But the tree has continued to gain ground. By 2008 it covered at least 43 acres on Mustang Island and two years later has started to grow on Galveston Island, 160 miles northeast.[4]

Recent DNA evidence suggests Hawaiian and Texan peppertrees descend from Floridian, not Brazilian, stock. Moreover, peppertrees in Florida are predominately *intra*specific hybrids—crosses between two different populations or variants of the same species—one introduced on the west coast of Florida (Punta Gorda) and the other on the east (Miami). Similar to the situation with *inter*specific hybrids of saltcedar, the extra genetic diversity may help pepperwood invade habitats not found in its homeland.[5]

Given the tree's track record in Florida and Hawaii, Texas officials took action. In 2014, the City of Port Aransas, Texas A&M Forest Service, and the Lady Bird Johnson Wildflower Center teamed up to form the Texas Gulf Coast Cooperative Weed Management Area. This management organization aims to remove, or at least control, peppertrees from Packery Channel on Mustang Island to Port O'Connor, including San José and Matagorda Islands—a 75-mile stretch of dune-filled barrier islands.

Brazilian peppertree has much in common with Chinese tallow. Rapid growth and aggressive spread allow it to crowd out and displace native plant communities. It evolves into nearly impenetrable tangles of pure pepperwood within a few years of invasion. Mammals (including feral hogs in Hawaii) and birds readily consume and disseminate its dense clusters of tiny fruits, having improved the germination rate among expelled seeds with the scarification of digestive-tract acids. Pepperwood invades marshlands and withstands flooding, salty soils, fire, and (unlike tallow) droughty conditions. One redeeming value the tree shares with tallow: its abundant nectar yields a peppery-tasting honey. Apiarists in Hawaii and Florida have favored the plant for this reason and, in the case of Florida, have opposed eradication programs.[6]

Pruning and cutting peppertree can cause contact dermatitis and edema. As the plant belongs to the cashew-mango family, which includes poison ivy and sumac, lengthy exposure to the plant's sap may result in lesions, oozing sores, and severe itching. Livestock that forage on the plant may be poisoned, and birds that feed excessively on ripe fruits are said to become intoxicated and unable to fly. The wood is useless.[7]

There are currently no effective biocontrol agents in use, though a sawfly, thrips, louse, and fungus have shown promise. Removing plants mechanically is effective, but care should be taken not to shake off its seeds. Larger trees can be bulldozed when not in fruit or girdled and sprayed with triclopyr, a foliar herbicide for woody brush and broadleaf weeds.

Giant Salvinia

Salvinia molesta is an aquatic fern, a tiny thing: just two leaves, each no bigger than a fingernail in its initial growth. Then the leaves on the water surface enlarge, fold up, and in crowded conditions become chainlike, sometimes forming mats as much as three feet thick. Like water hyacinth, giant salvinia spreads vegetatively. Some botanists think the world's entire population comes from a single genetic individual.[8]

Giant salvinia at various stages of maturity: (upper left) young plants with leaves floating on water's surface; (center) mature plants, compressed and crinkled. Photo by Eric Guinther (Marshman in Wikipedia) (CC BY-SA 3.0).

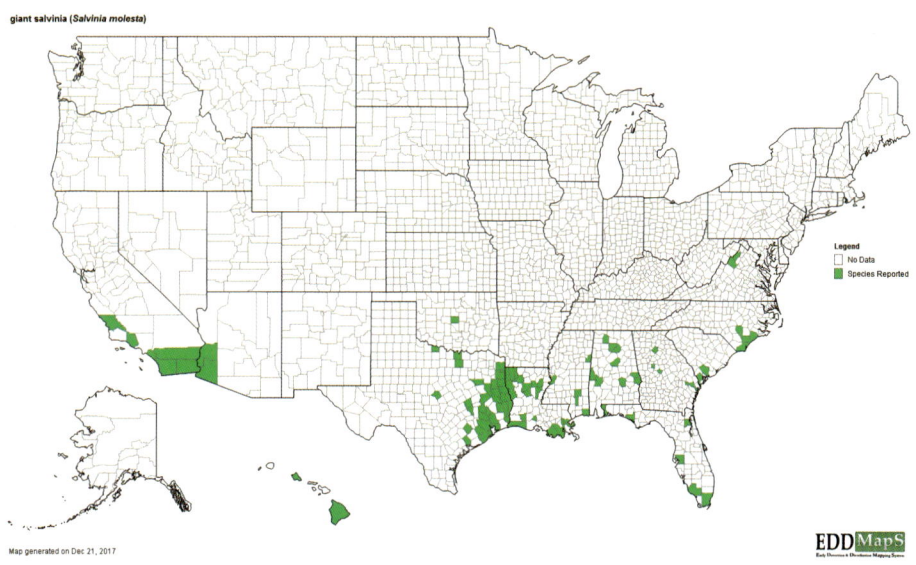

US distribution of giant salvinia (*Salvinia molesta*). EDDMapS, University of Georgia—Center for Invasive Species and Ecosystem Health, 2017, http://www.eddmaps.org/.

What is alarming about giant salvinia is its astounding growth. In laboratory conditions, the number of plants can double in two to four *days*. In the field, doubling takes only a week, twice as fast as water hyacinth's rate of increase. Some experts claim salvinia puts on biomass five times faster than water hyacinth. In fact, calculations suggest a single plant can cover 40 square miles in only three months.[9] The threat of all this growth sounds familiar. Water-intake pipes, ditches, channels, and canals become clogged; wildlife, livestock, and anglers find access to water bodies difficult; navigation is curtailed; and mosquito habitat is expanded.

Native to southeastern Brazil, where it neither is especially abundant nor forms extensive mats because native insects keep it in check, giant salvinia is now found worldwide because of its pleasing look in aquaria. Some plants arrived in Sri Lanka in 1939 and then spread into India, Southeast Asia, Oceania, the Caribbean, and Africa.[10]

Giant salvinia first hit the United States in 1995 when it turned up near a 1.5-acre pond in Walterboro, South Carolina, from which it was eradicated. But three years later, it surreptitiously appeared in a schoolyard wetland demonstration pond in Houston, Texas. That same year it popped up at several other sites in Texas—such as Toledo Bend Reservoir—as well as in Louisiana, which borders that impoundment. By 1999, dense infestations of the weed were growing in the lower Trinity River and in many sites in the US Southeast, Oklahoma, Arizona, California, and even in Hawaii.[11]

All these appearances happened despite salvinia being listed as a federal noxious weed, which means people are prohibited from bringing it into the United States and transporting it across state lines. The fern had likely been an aquarium staple for years, and inevitably it got into the wild, where fragments from boats and boat trailers facilitated its spread.

Control techniques are vexing. Mechanical removal is expensive and ineffective given that fragments simply resprout. Chemical control, while effective in small, contained sites, is generally impractical for large-scale projects. Herbicide sprays are not always effective, as salvinia stems and buds grow underwater and its leaves are densely covered with hairs that cause chemicals to bead up and roll off.

Caddo Lake, the largest natural lake in Texas, is an important case in point. The 25,000-acre lake bordering Louisiana features one of the largest bald cypress forests in the world and is Texas' only Ramsar Site, meaning it is a wetland of international importance, one of only 38 in the nation. Giant salvinia appeared in Caddo Lake in 2006 and within a few weeks covered 30 acres. "Weed warrior" volunteers pulled out plants by hand, and park officials sprayed herbicide costing $1.5 million over the remainder. Authorities tried burning the plant, slinging booms and barricades, and employing mechanical harvesters. None of these efforts worked. By 2012, the floating fern covered almost one-half of the Texas side of the lake.[12]

As in the situation with water hyacinth, a tiny weevil has come to the rescue. *Cyrtobagous salviniae*, commonly known as the salvinia weevil, originates from southeastern Brazil, where its larvae feed solely on salvinia buds and rhizomes. After infestation, the ferns turn brown, become waterlogged, and sink.

At first, weevil results were mixed, because the insects are not cold tolerant. The Morley Hudson Weevil Greenhouse at Caddo Lake, "the first high-production weevil greenhouse facility in the world," was established to assist with research.[13] Experts there helped identify weevil strains that worked best in East Texas. Success came in 2013 after weevils cleared 10–15 acres at one site. Recent work suggests that combining weevils and sprays gives the best results.[14]

At last count, giant salvinia is growing in more than two dozen water bodies in Texas, including Lake Fork (another of the state's top bass fisheries), Martin Lake, Lake Murvaul, Pinkston Reservoir, Lake Sam Rayburn, and Lake Livingston. It is infesting water bodies around Houston. Experts hope the weevil will continue to be effective, though Hurricane Harvey's devastating flooding has likely spread plant bits more widely in Southeast Texas.

UGA1317057

The salvinia weevil holds promise as a biocontrol agent. Note the distinctive (and diagnostic) egg-beater-shaped hairs on giant salvinia's leaves. Photo by Scott Bauer, US Department of Agriculture, Agricultural Research Service, Bugwood.org.

Zebra Mussel

**The Zebra Mussel (*Dreissena polymorpha*) has become the most
serious nonindigenous bio-fouling pest ever to be introduced into
North American freshwater systems.**
—US Army Corps of Engineers, 2002

We have discussed several invasive plants that infest our reservoirs, but
there are animals, too. One recent and insidious invader, the zebra mussel,
a small striped shellfish 1.5–2.0 inches long, is busily floating its way across
Texas after appearing in Lake Texoma in 2009. It is causing significant economic and ecological harm to human and biological communities.

The sharp-edged freshwater mussel is reproducing in at least 14 lakes
and 5 watersheds; it has been detected more than once in 5 additional lakes
and once in 3 more. In June 2017, several shellfish were retrieved from Lake
Travis, one of the Highland Lakes in Central Texas, and two months later
half-inch specimens came from around Tom Miller Dam, which holds Lake
Austin downstream from Lake Travis and is the spillway for Lady Bird Lake
in downtown Austin. Texas Parks and Wildlife officials speculated that one
or more vessels had carried them from another water body, perhaps Lake

Zebra mussels are quite
small but grow in exceed-
ingly dense populations.
Photo by Matt Turner.

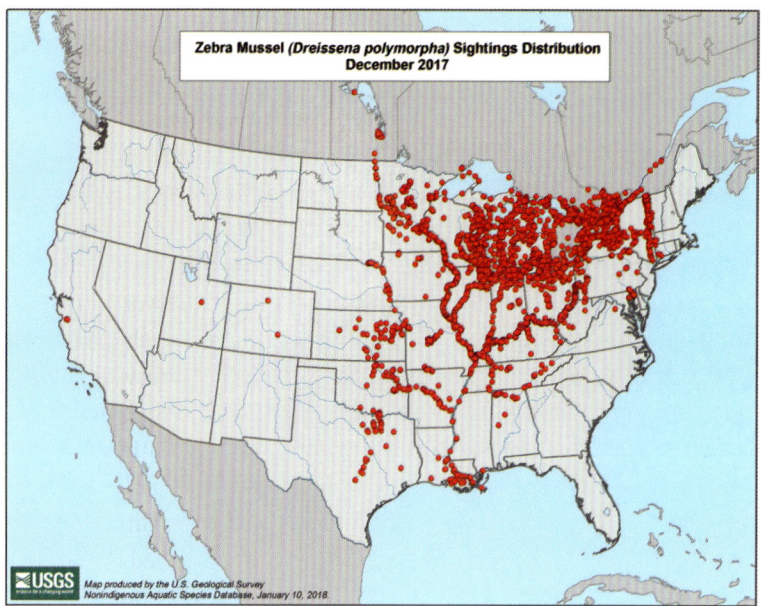

US distribution of zebra mussel sightings. In less than a decade the small shellfish have spread in Texas from Lake Texoma, on the Oklahoma border, to at least 20 water bodies along several major river watersheds. Photo by US Geological Survey, Nonindigenous Aquatic Species Database—Gainesville, Florida, 2017, https://nas.er.usgs.gov.

Travis, which is infested. In eight years, the mussel has found its way into 20 water bodies in the state, and the tally will grow.[15]

Lake Texoma is 75 miles north of Dallas and about 6,200 miles west of the shellfish's original home in the Black Sea and Caspian Sea. The minuscule zebra mussel reached that reservoir on the Oklahoma border after its ancestors had clung to a ship's hull or were carried as microscopic larvae in ballast water and pumped perhaps into Lake St. Clair, connecting Lake Huron with Lake Erie on the Michigan-Ontario border, where they were first identified in 1988. As a female can release 30,000–40,000 eggs in her first year, shellfish larvae quickly drifted through this largest set of lakes on the planet. By sticking to hulls, outboard engines, and boat equipment (the shellfish survives out of water for a time), or simply as larvae, zebra mussels infested the Mississippi River and its tributaries all the way to Texas.

Lake Texoma once supplied almost 28 billion gallons of water annually to customers in the North Texas Municipal Water District. However, once the shellfish was declared a hazard, the US Army Corps of Engineers shut down the flow to prevent the invasive species from entering the Trinity River (it did, however) and made district engineers construct a new 46-mile-long pipe to supply their customers. It cost $300 million, causing water rates in Tarrant County to jump 14 percent. Houston residents, who rely on Lake Livingston, recently infested with the mussels, took note of this expensive ploy.[16]

The zebra mussel is a filter feeder that clings to underwater structures, feeds avidly, and reproduces easily. Huge clumps build up on piers and beams, clog outlet and intake pipes, and can immobilize native shellfish by clinging to them. The sharp shells are not kind to bare feet or to swimmers bumping into them on piers. Populations grow so massively that they can deplete the dissolved oxygen in a water body, essential for native clams, fish, and plants. An adult-sized mussel pumps about two pints of water a day through its open shell, filtering nutrients and expelling fecal matter. It feeds efficiently, generates waste, and may reduce food supplies needed for bait and game fish desired by anglers. The filter feeder clarifies the water, often to the delight of lakeside residents, but more light in the water column alters native fish and plant habitats. So profligate did some populations become in the Great Lakes that outbreaks of avian botulism have been

Zebra mussels infest all types of submerged objects, including this discarded shopping cart. Photo by Lames F. Lubner, University of Wisconsin Sea Grant Institute, Bugwood.org.

associated with them due to the concentration of bacterial toxins passed to waterfowl that consumed them.

Currently Texas Parks and Wildlife's "clean, drain, and dry" policy requires boaters to check before and after launching. The state's aim is to form a coalition of fishers and boaters that reinforces a practice to stop this invasive organism from spreading. Control involves hosing off boats to remove clinging mussels. Applying chlorine-based chemicals also kills them.

One promising control is the use of Zequanox (developed in 2007), a bacterium-based molluscicide that is reportedly benign to native clams. After testing more than 700 bacterial strains, in 1995 Daniel P. Molloy discovered a strain of *Pseudomonas fluorescens* that killed zebra mussels, and the equally invasive quagga mussels, by being absorbed into and then destroying a shellfish's digestive tract. Promoted as a biocontrol, the substance is expected to be economically viable for use in specific sites and situations.[17]

Lionfish: Hogs of the Sea

Melissa Gaskill asks an odd question: Can we eat our way out of an invasive species problem? She is referring to an attractive, edible fish about 15 inches long that weighs about two pounds. The problem is that the home of the red lionfish is in the western Pacific Ocean, so this gaudy "feathered flutterer" (an anglicized approximation of its scientific name, *Pterois volitans*) has been multiplying in its new home in the Gulf of Mexico and Caribbean Sea. An adult lionfish has recently been speared on a reef in the South Atlantic Ocean, about 5,000 miles southeast of Galveston.

Pests have been on the menu before. Nutria, the invasive rodent from South America, which inhabits the eastern two-thirds of Texas, was promoted at game fairs and even in the fancier restaurants in Louisiana as good to eat. But the promotion failed. People apparently did not take to eating a giant water rat, despite meat that was lean and low in cholesterol. Now, it is a fish's turn, although one has to snip off 13 long venomous dorsal and 3 anal spines to make it safe to handle.

Red lionfish are stunningly beautiful, and it is easy to see why aquarium keepers collect them. The problem is they do not stay inside glass walls forever. Melissa Gaskill describes how the Florida pet trade (assisted by tropical storms) was probably the mid-1980s source for this omnivorous, slow-moving tropical species that suddenly lunges at native fish and vacuums up eggs, fry, and most anything it can ingest whole on coral reefs and beyond.[18] At least 70 fish and invertebrate species have fallen to lionfish, which can distend their

Red lionfish. Since its appearance along Florida's coast, the fish has steadily spread up the Eastern Seaboard and southward into the Gulf of Mexico and Caribbean Sea. Predicted future expansion includes much of the Atlantic Coast of South America. Photo by Symbiosus.

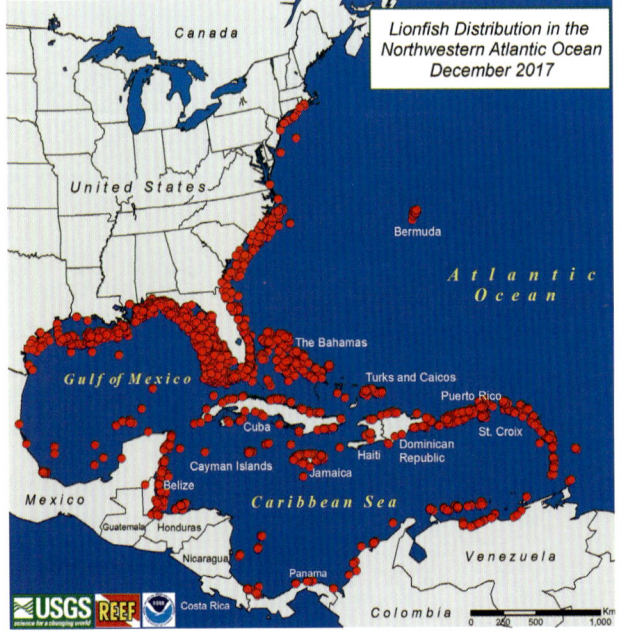

Lionfish distribution in the northwestern Atlantic Ocean in 2017. Photo by US Geological Survey, Nonindigenous Aquatic Species Database—Gainesville, Florida, 2017, https://nas

.er.usgs.gov.

stomachs to carry anything that they can fit into their gaping mouths. Free of the large grouper, shark, and eel predators of its native range, the lionfish has become explosively abundant in new waters. Like another pet, the havoc-causing Burmese python in the Florida Everglades that places coastal Texas in its future range, lionfish owners discarded or released specimens into warm waters off eastern Florida. Unfortunately, it has arrived here.

Lionfish turned up on the Flower Garden Banks, named after colorful native sponges, a protected coral reef system about 100 miles off Galveston in 2011. It is the lone national marine sanctuary in the Gulf of Mexico, characterized as a "Gulf Treasure" filled with dense, healthy, northern-living corals, resembling "a Caribbean oasis in an unexpected place." Currently, recreational divers see more and more lionfish as the fish feed in the morning and evening over the reefs. In 2011, scuba divers counted 17 fish in three sections of the banks; in 2015, the count climbed to 2,614.[19]

In her article, Gaskill refers to the lionfish's "roar" increasing in Texas waters. We prefer to compare it to the squeal of the feral pig because the invasive lionfish is proving a real hog among native fish—adaptable, aggressive, and fecund. She refers to Keri Kenning of the Reef Environmental Educational Foundation as noting it likes artificial as well as natural reefs, mangroves, and seagrass flats (and even pushes into rivers), and it does well in waters much deeper than a scuba diver's range. These scorpionfish devour more than 50 native fish species as well as their eggs, including juvenile snapper and grouper, and are growing larger than Pacific Ocean kin. There is no end in sight, as a single female is capable of laying two million eggs annually and may live for 15 years in her native range.[20]

Experts argue that lionfish derbies are one way of denting populations, provided fishers hold them regularly. Once speared and the venomous spines clipped off, the fish may be cooked in interesting ways, as listed in *Lionfish Cookbook*.[21] The fish's meat is deliciously light and flaky, and enthusiasts can kill and eat as many as they wish to save a local reef or native fish spawning site. The idea is to coordinate activities so that fishers can get them first before they get us, says Leslie Hartman, with Texas Parks and Wildlife, which is partnering to develop a management plan. It includes promoting lionfish as delicacies and deploying underwater robots to hunt the fish that swim below the 90-foot depth limit for most recreational diving.[22]

A 3.5-foot-long camera-mounted robot that is directed from the surface to a depth of 400 feet was promoted by the British contingent for the America's Cup held in Bermuda in 2017. The robot creeps up on the fish, hits it with two electrical probes, and then stows it and goes after another until it surfaces with these fish-hogs. America's Cup boaters agreed to keep marine waters clean and protect habitats around Bermuda, a favorite lionfish locale. This meant going after lionfish, which are stripping island waters, and making the robot cost effective among local fishers.[23] However, as in the case of the feral hog, eradication is highly unlikely, and even substantial reduction will always depend on habitual and organized efforts to catch them. The lionfish, like the hog, is here to stay.[24]

Emerald Ash Borer

> It's safe to say that the vast majority of ashes [in North America] will surely die.
> —Andrew Liebhold, research entomologist, US Forest Service

The adult emerald ash borer is so beautiful it would befit a brooch or bracelet. The half-inch-long beetle is a bright metallic-green, with iridescent hints of copper, gold, and azure. Beneath its forewings, the dorsal surface of its abdomen is a brilliant coppery-red. This is a feature that separates *Agrilus planipennis*, native to eastern Asia, from the 170 other species of this insect genus found in North America. While few Texans have seen this member of the aptly named jewel beetle family, we are all likely to see evidence of its destructive force.[25]

The emerald ash borer feeds on any species of ash tree (*Fraxinus*). During its larval stage, which may last up to two years, the borer is a white, 1.5-inch worm that depends exclusively on the tree's water and nutrient layers beneath the outer bark. The larva carves telltale serpentine tunnels or galleries, which are packed tightly with its sawdustlike waste, called frass. In late spring and early summer, adult beetles bore diagnostic (one-eighth-inch) D-shaped holes through the outer bark, where they emerge, mate, and search for other ash trees so they can lay eggs on their bark and nibble their foliage.[26]

Eating leaves does little damage to the ash, but tunnels made by myriads of hungry larvae kill the tree by girdling it. The tree's upper canopy thins; then whole branches die. Woodpeckers take special interest and peck out larvae. The bark sometimes splits and exposes underlying galleries. As the tree fights to survive, it starts sprouting aggressively from its trunk, creating shrublike growth at its base or midsection. This last-ditch effort, called epicormic sprouting, is rarely successful in keeping the tree alive. Most ash trees die two to three years after infestation.

Hundreds of millions of ash trees in North America have succumbed to the emerald ash borer. Likely introduced via packing materials, such as wood pallets and crates, into southeastern Michigan in the 1990s, the borer was not identified until 2002 after it had already killed thousands of trees. Since mated female beetles usually fly less than two miles from their site of emergence before laying eggs, it is humankind who once again has assisted the invasive

Emerald ash borer. The tiny beetle has so far appeared in only one county in East Texas, but future expansion in the state is likely. Photo by James E. Zablotny, US Department of Agriculture, APHIS (CC BY 2.0).

An emerald ash borer beetle and the frass-filled serpentine tunnels just below the tree's bark that the beetle digs during its larval stage. Photo by Eric R. Day, Virginia Polytechnic Institute and State University, Bugwood.org.

insect's dispersion. By moving infested firewood, nursery stock, green lumber, and wood chips, we have unknowingly helped the beetle march across the entire eastern half of the United States into 31 states and two Canadian provinces. There are said to be 7.5 billion ash trees nationally in rural areas, and millions more in urban ones. Cities in the Midwest, where ash makes up 20 to 80 percent of urban and suburban forests, have experienced massive die-offs. Fort Wayne, Indiana, for instance, lost 14,000 ash trees on city streets alone. National cost estimates for the treatment, removal, and replacement of trees infested by the borer in the United States just through 2020 approach $13 billion, leading forest researchers to call the beetle "the most destructive and costliest forest insect to invade North America to date."[27]

It is a sad irony that many neighborhoods had planted ash trees to replace the American elms that had been killed off in the mid-twentieth century by the exotic fungus known as Dutch elm disease, which entered the country on a shipment of logs for a furniture company and was spread by both native and nonnative elm bark beetles. With both the ash and elm gone, city planners and some homeowners admitted they had relied too heavily on one tree, thereby creating a woody monoculture.

Texans knew the emerald ash borer was headed into the state. Texas A&M Forest Service personnel began setting out insect traps in 2012. The beetle turned up in Colorado in 2013, in Arkansas the following year, and in Louisiana the year after that. Officials were probably not surprised when in April 2016 four adult beetles were confirmed in a monitoring trap just south of Karnack, Texas, in Harrison County six miles west of the Louisiana border. In 2017 five more were found at the same site. State and federal officials continually check this area for affected trees and are deploying additional traps. To date, no dying trees have been found, but it is only a matter of time.[28]

In its homeland in northeastern China, Mongolia, the Russian Far East, Korea, and Japan, the borer is not a significant pest because predators, such as birds and wasps, take a toll, and local ash trees have developed a natural resistance to infestation. Outside its native range, of course, the story is different. So far, biologists have found no natural resistance among the 16 species of ash thought to be threatened in the United States, and mortality is nearly 100 percent in heavily infested areas. The beetle was believed to affect only ash, but in 2014 it began to attack the white fringetree (*Chionanthus virginicus*), an understory ornamental of the eastern deciduous forest that belongs to the same family as ash.[29]

There are eight species of ash native to Texas. The most common ones are green ash (*F. pennsylvanica*) and white ash (*F. americana*) in the eastern half of the state and Texas ash (*F. albicans*) located in central counties. Foresters estimate that ash species make up less than 5 percent of Texas forestland. However, that percentage means millions of individuals. In urban areas, developers have extensively planted two quick-growing ashes for shade: Arizona ash (*F. velutina*) from the Trans-Pecos and Mexican ash (*F. berlandieriana*) from South Texas. Heavy reliance on these short-lived, insect-prone species—both marketed as Arizona ash—will likely lead to large die-offs and cost homeowners to have big specimens removed.

Controlling the emerald ash borer is hard. Systemic insecticides help, but they work best before the tree is infected or shortly thereafter. Treatment is a long-term affair, with repeated applications to the trunk and surrounding soil. Arborists recommend waiting for the beetles to be discovered in the general area and then saving insecticides for specimen trees. Three species of parasitoid wasps, which specialize in attacking the borer's eggs or larvae, are being released in the Midwest for biocontrol. Some have shown initial signs of success, but like all biocontrol agents, they will help slow the spread but not stop it. Harsh as it may be, some experts recommend cutting down ash trees, especially those larger than 10 inches in diameter, and replacing them with other tree species, including ash hybrids that are resistant to the borer, as these are developed and become available. Some communities have elected to chop down their ashes ahead of time so they can schedule plantings and spread out the costs. Researchers will be on the lookout for native trees that survive the borer onslaught, as these can lead to future strains of resistant trees, as happened with the American elm.[30]

Two simple lessons can be learned that help slow the spread of this invasive beetle, as well as other insects and pathogens: purchase firewood locally, or "buy it where you burn it," as the Nature Conservancy says; and always plant a variety of tree species in urban areas.[31]

White-Nose Syndrome in Texas Bats

> Today, the US Fish and Wildlife Service announced over $1 million in grants to 37 states and the District of Columbia to help combat white-nose syndrome (WNS). . . . The grants bring the total funding to states for WNS response over the last eight years to $7 million.
> —US Fish & Wildlife Service, July 2017

It took merely 10 years for the fungus that causes white nose syndrome (WNS) to make it from the state of New York to the state of Texas. Since

Little brown bat exhibiting the fungus that causes white-nose syndrome. By 2018 the fungus has been identified in four species of bats across eight counties in the Texas Panhandle and two counties in the Hill Country. Photo by Ryan von Linden, US Fish and Wildlife Service.

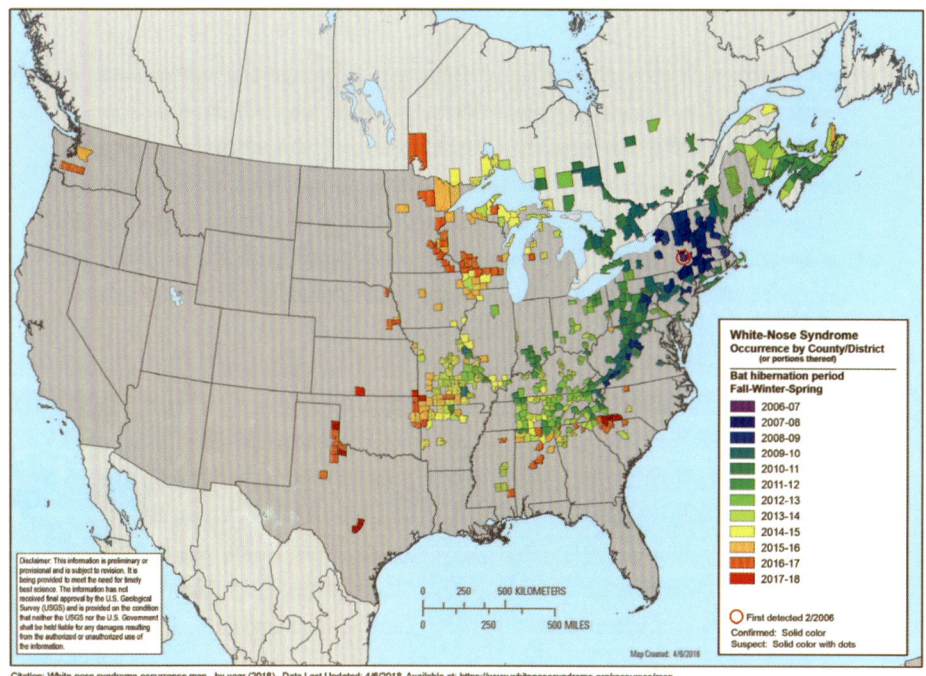

White-nose syndrome occurrence map, April 2018. US Fish and Wildlife Service.

2006, when white-nose syndrome was first detected in four caves west of Albany, New York, about six million bats have died in North America from this disease caused by a fungus, *Pseudogymnoascus destructans*. The fungus appears on the muzzles, ears, and wings as white cottony blotches and springs up in caves where bats hibernate. The disease drains an individual's fat reserves and causes it to awaken and seek food in wintry conditions. Unable to find sustenance, the bat weakens and dies of starvation.

Unfortunately, the colder, exposed Panhandle made Texas the thirty-fourth state to report the fungus, of which, together with five Canadian provinces, have reported WNS. Surveys in 2017 detected three species of bats carrying the fungus in six Texas Panhandle counties, though none of them exhibited symptoms of WNS. More ominously, in April 2018, Texas Park and Wildlife personnel discovered bats in Central Texas with the fungus, including a single Brazilian, commonly referred to as Mexican, free-tailed bat (*Tadarida brasiliensis*), a highly migratory and numerous species that attracts thousands of onlookers to Austin's Anne W. Richards Congress Avenue Bridge on most summer evenings to view the exodus of hungry bat mothers.

The bat in question came from Old Tunnel State Park west of Austin and south of Fredericksburg. The fungus has turned up in nine additional counties. This was the first time officials had noted this bat species with the white-nose fungus; and as bats in general and Mexican freetails in particular control insects to the tune of $1.4 billion annually, the repercussions to Texas' agriculture as well as to its 32 bat species are potentially profound.[32]

As Mexican free-tailed bats migrate rather than hibernate, experts are worried because the bat does associate with others that do, including species known to have WNS. So Texas Parks and Wildlife personnel are concerned whether this humanly spread disease, which occurs in Europe, where it is not virulent to native bats, will decimate millions of bat cousins in Texas. Somehow the fungus crossed to North America, perhaps on cave visitors' shoes or equipment, and infected native bats. Control efforts listed in a TPWD *Action Plan* include responding quickly to outbreaks, monitoring impacts, and preventing spread. The latter challenge includes curtailing access to bat caves wherever possible, using cave decontamination and bat-handling protocols so that cavers and researchers do not carry and spread fungal spores.[33]

Nationally seven bats species (two of which occur in Texas) are known to have contracted WNS, and a similar number of species (all are in Texas) have tested positive for the fungus without manifesting the syndrome. Ongoing research includes using US bacteria in cave soils to inhibit the growth of the fungus, which does best in very humid conditions between 40°F and 55°F. Exposing infected bats to aerosols containing a common bacterium has been shown to improve their health and inhibit the spread of WNS. However, the possibility of reinfection after release remains. Efforts are also afoot for so-called gene silencing, whereby the genes that direct a certain expression are knocked out so that the syndrome may eventually become harmless.[34]

Conclusion

Biological pollution, in the form of exotic species, is now one of the leading threats to the ecological integrity of our forests, grasslands, and waterways.

—Bruce A. Stein and Stephanie R. Flack, *America's Least Wanted*[1]

The nonnative animals and plants in this book are a representative selection of the most serious invasive species in Texas: they are abundant, widespread, tenacious, and damaging. Various methods have been taken for dealing with them, although most experts agree it is unlikely if not impossible to eradicate them. Their stories, though superficially similar, are nuanced and ultimately idiosyncratic and reflect a variety of beliefs, assumptions, attitudes, and feelings that come into play with nonnative species. They complicate simplistic binaries of native versus exotic, good versus bad, preservation versus eradication.

Lessons Learned about Texas Invasive Species

In one way or another, people are responsible for importing all the invasive plants and animals we discuss. In most cases they were released intentionally under an unquestioned belief that they were important in making Texas a secure and satisfying home. Directly or indirectly, we are agents of dispersing them. The discussions have pointed to the need for more care about how and why we do this. Neither starlings nor the much more likable monk parakeets flew here on their own. Axis deer did not stray over our border in the way the nine-banded armadillo appears to have done; hydrilla tubers did not wash ashore, nor did saltcedar seeds waft down to earth. Until recently, we did not challenge the assumption that shifting and mixing flora and fauna were ways of improving Texas, even though they have affected the biological integrity of the state.

In some cases, plant and animal introductions were accidental, but we are still responsible for them. On cargo ships, red imported fire ants hitched a ride in soil ballast, while the microscopic larvae of the zebra mussel arrived in water ballast and has taken boat rides from its port of entry to reach

Texas. The emerald ash borer stowed away in packing crates that bore commercial goods. A fungus that decimated the American elm spread on lumber destined for furniture. Another fungus that is wreaking havoc on native bats appears to have spread through caving equipment, including spelunkers' boots, and the bats infected by it have now passed it on as far as Texas. In this age of ever more voluminous and speedier transportation, unintended introductions are only likely to increase in the Lone Star State as well as around the world. This book is a call for vigilance.

We are continuously reshaping the natural environment with opportunities for both native and nonnative organisms to mingle and adapt. Cattle drives, fencing, and fire suppression allowed native mesquite to spread across much of the state. Feedlots in the Panhandle provide handouts for alien starlings, while feeders for deer help nonnative hogs thrive. The multiplication of suburban lawns, city parks, and golf courses has attracted native great-tailed grackles and recently white-winged doves. Transmission towers overlooking malls and schoolyards are being colonized by nonnative monk parakeets. Damming a river creates downstream habitats that introduced saltcedar may exploit, while above the dam exotic water hyacinth, hydrilla, or giant salvinia may inch across the lake.

Introducing an alien plant or animal, or subspecies of it, inadvertently allows it to spread beyond its native range and perhaps hybridize. For instance, more than 80 percent of the saltcedar in the United States consists of hybrids between a saltcedar from eastern China and another that extends westward as far as Turkey. The Brazilian peppertree that is now proliferating in coastal Texas did not arrive from its home in Brazil but from Florida, where two different populations or variants of the same peppertree have crossed to form a so-called intraspecific hybrid. Experts often reason that hybrid vigor enhances an organism's chances for success in a new habitat. Red and black imported fire ants have hybridized and are well established in southern states. Animal breeders have crossed goat species to obtain the Texas hybrid ibex, which never sheds its horns; different breeds of sheep form the Corsican ram; and antelopes interbreed to produce so-called scimboks, stunningly colored and often bearing longer horns than their oryx species parents. These new trophy animals may produce fertile offspring that adapt to drier Texas lands.

And, of course, we are the ones who carefully import exotic organisms to attack invasive species. Biocontrol involves selection, testing, rearing, and release: Kazakhstani and Tunisian beetles devour saltcedar; Argentine flies and moths consume water hyacinth; Brazilian weevils tackle giant salvinia; South American phorid flies parasitize fire ants; East Asian grass carp chomp on hydrilla. Biocontrol organisms have largely performed well. But when things run amuck—beetles spreading faster than expected, winds blowing moths into neighboring areas, fish devouring native plants as well as exotic weeds—we take stock and reassess methods of control.

We belabor the human role because we quickly forget it. In fact, there is a tendency to blame the animal or plant for being here and doing what it does. As already noted, the word "invasive" conjures up armies that are targeting us, conveniently concealing the fact that the responsibility for the invasion belongs to us. By definition, all nonnative species are ones that humans have introduced. Thus, the invasive moniker, which connotes harm (usually to humans), points out how well these nonnative species are flourishing. Invasive species are simply doing what all living things do—feeding and reproducing—most abundantly in propitious habitats, without predators, diseases, and other limiting factors. It is easy to lose sight of this amid the sensational epithets we tend to use when talking about them, such as "Top 10 Worst Invaders," "Monster Plants," "World's Worst Weeds," and "Zombies at Your Door," and so on. We are the ground zero here.

As we have seen, only a tiny percentage of the organisms we move flourish in new places. Most do not survive, but for various reasons some do. These successful organisms exist in a kind of ecological nirvana; everything is working in their favor, although it may take time before their biological exuberance and success become apparent.

In the case of invasive species, this exuberance ultimately implies harm. The house sparrows' adaptability made them one of the most populous birds in North America, overwhelming native species in their quest for nest spaces, threatening purple martins in particular. Cats are amazingly efficient killers when allowed unfettered access to outdoor wildlife. Hogs turn into "Sherman tanks" in a feral state, leaving croplands looking more like battlefields. Water hyacinth, giant salvinia, and hydrilla, having gained access to rivers, lakes, and bayous, impede navigation, damage water structures, push out native plants, interfere with recreation, and expand mosquito habitat. The lovely Chinese tallow is the greatest woody threat to the coastal prairies in Texas. Saltcedar reduces the diversity of, and intensifies hydrological changes in, our riparian corridors. Red imported fire ants infest millions of acres of crops, pastures, parks, and suburban neighborhoods, causing more than $1 billion in damage in Texas every year. Exotic game is turning the Lone Star State into a "huge outdoor laboratory" of escapees and hybrids that diminish available forage, compete with native animals, and introduce new diseases. Lionfish are clearing offshore reefs of native fish larvae. The emerald ash borer is poised to wipe out eight species of ash tree in Texas, while white-nose syndrome is on the verge of spreading among bat populations. Brazilian peppertree is encroaching on our coast, threatening a repeat of its invasions of Florida and Hawaii. Historically, we have ignored or downplayed the costs of introducing organisms like these. And we have dealt with only a select few, known to be the worst, in this book. The full list is long, is continuing to grow, and makes depressing reading.

But it is not all doom and gloom. There are upsides even among the most vexing species, which we are understandably reluctant to admit in our zeal to deal with them. All these invasive organisms are part of nature, of course, and they function within ecosystems by absorbing water, emitting oxygen, cycling nutrients, controlling pest outbreaks, and so on. When these basic functions provide benefits to people, they are called ecosystem services. Such services have become a hot topic in ecological circles because they are ubiquitous and important, but we have often tended to overlook or undervalue them. For instance, plants produce oxygen while sequestering carbon in their roots; the former allows us to breathe, while the latter ultimately keeps the world cooler. Pollinators are important to native plants, which could not reproduce without them, but agriculture depends on them as well. Many water plants are excellent filters of water and make perfect shelters for fish. Birds and bats check insect populations, which may lower certain crop pests and reduce diseases. The list is enormous since it describes nothing less than how the intricate web of life functions.

It should come as no surprise, therefore, that invasive species also provide such services. Hydrilla is floating cover for largemouth bass in artificial lakes, so anglers benefit. Chinese tallow produce fruits on which 35 species of US birds feed. Tallow trees also produce copious amounts of nectar, as do eucalyptus and Brazilian peppertrees, which honeybees (another exotic creature) gather. Beekeepers have resisted having the tallow listed as a noxious weed. Saltcedar offers windbreaks, erosion control, shade, and nesting sites for birds. Raising exotic game bolsters populations of threatened animals and possibly preserves some from extinction, as well as provides employment and revenue. Cats, which comfort us as pets, help control rodents. Water hyacinth, "the most troublesome weed in the world,"[2] helps in wastewater treatment and checks the spread of green algae. We all know hogs yield bacon and ham, but we should also accept that nutria, lionfish, and grass carp are perfectly edible too, even delicious. These benefits complicate but also enrich our attitudes toward management strategies.

Attitudes and emotions also affect the invasion issue. At one time we preferred hogs to native javelinas, tomcats to bobcats, foreign lookalike sparrows to native sparrows, saltcedars to local willows and cottonwoods. Conventional wisdom assumed that crop plants, domestic livestock, and pets, though mostly alien species in North America, were easier to manage than native plants and animals and much more useful. And when these alien species have been with us for a long while, we consider them native or as family. Many Californians embrace the eucalyptus tree, for example, just as most Americans accept the house sparrow and European starling, so abundant and conspicuous in our towns and cities, as native birds, largely ignoring or downplaying the damage they do. Our feelings about feral cats and feral hogs, for instance, could not be more at odds, and moral issues are raised over no-kill policies for cats but not the hogs. Exotic game casts into sharp relief our desire to preserve attractive, graceful animals, some of which are endangered in their homeland, with our desire to hunt them, and enjoy the cash this generates. Our love of the animals, our respect for the hunt, and our pride in saving rare and threatened species may help us ignore whatever impact these exotics are having on native wildlife and Central Texas landscapes.

Usefulness has driven our desire for bringing in exotic species. In fact, the federal government actively sought out potentially useful plants through the Foreign Plant Introduction Division of the USDA. But almost everything is useful for something in a time and place. Moving a species into a new area changes it in some way. House sparrows were assumed to be insect eaters in the United Kingdom and were imported to rid us of inchworms in our noisy industrializing cities from which native birds were fleeing. Instead, they shunned worms for grains. Benjamin Franklin promoted Chinese tallow for its oil, but two and half centuries later the tallow industry has yet to be born, despite the fact that the tree is a very productive source of oil. The promoters of saltcedar claimed it tolerated salty conditions and drought and helped control riverbank erosion while affording excellent shade. It did all these things quite well, but no one realized that such adaptability helped it outcompete native woody plants and spread far and wide across river valleys. Nutria offered new fur products to a thriving export industry until the whims and fad of fashion shifted and the market for pelts crashed, leaving burgeoning colonies of the rodents to consume wetland vegetation.

People agree that beauty enriches lives, so some noxious species arrived because people believed them beautiful as well as materially useful. Chinaberry held promise for both timber and fuel, but its value as an orna-

Many well-known landscape plants, such as (*left*) mimosa or silk tree (*Albizia julibrissin*) and (*right*) heavenly bamboo (*Nandina domestica*), were introduced for their ornamental appeal. Their beauty encourages their propagation and tends to keep them off state invasive species lists even when they have become naturalized and spread quickly. Photos by Charles T. Bryson, US Department of Agriculture, Agricultural Research Service, Bugwood.org, and Matt Turner, respectively.

mental shade tree with fragrant lilac flowers resulted in its spread after the 1790s. Saltcedar arrived on the East Coast in the early 1800s as a botanic garden curiosity, and nursery catalogs promoted it mainly for its looks; only later did officials take to it for erosion control and arid land planting. Heavenly bamboo, European and Chinese privets, mimosa or silk tree, Russian olive, tree of heaven, and Brazilian peppertree—the list goes on—held strong appeal because of their appearance, adding novelty, color, or shape to a landscape, park, garden, or street. Some of our most weedy aquatic plants and animals, such as water hyacinth, hydrilla, giant salvinia, and the red lionfish, got their start because of their beauty, grace, and color in ponds and aquaria. Cats are generally considered graceful and lithe, even among people who are not fond of them. And at least some of the mammals imported from Asia and Africa, such as the axis deer, blackbuck antelope, and scimitar-horned oryx, are stunningly elegant and carry spectacular horns, which clearly add to their appeal.

What Can We Do?

In an ideal world, we would neither import nor distribute nonnative species, or at least be proactive and take great care about doing so. Our natural environment evolved its complex communities of plant and animal interdependencies over long periods of time. Although this web of connectivity is constantly changing to reflect environmental shifts (such as drought, fire, or now climate change), the addition of nonnative animals is likely to disrupt existing ecosystems. At the practical level we should pay more attention to the native species in regions of the United States, parks, schoolyards, roadsides, and similar public places, and in promoting them in our homeowner associations, plant nurseries, and garden outlets.

In the real world, organisms are transported worldwide to places where they do not occur naturally. We have surrounded ourselves with exotic species. What can we do? There is usually a window of opportunity after the first reports of a new organism doing damage. It happened with the zebra mussel, but reaction time took too long to deal with it effectively. The lesson is that when a nonnative organism known to be invasive turns up, or an alien one shows signs of becoming invasive, it is time to act. We are getting better at this, knowing that this is possibly the best opportunity for eradication. People who spend time outdoors and notice and then report a strange plant, or bug, or diseased bush or tree are a first line of defense. Field guides and manuals that identify invasive plants and insects are increasingly available to citizen volunteers, who work in national wildlife refuges, state parks, and city preserves to combat invader species. Organized and educated, they are a second line of defense. For example, TexasInvasives.org publishes eco alerts on a regional basis and encourages citizens to sign up for breaking news about invasive species. Its website encourages users to log in and report the presence of an invader. Partnered by nine agencies and institu-

Spotted knapweed. Photo by Rob Routledge, Sault College, Bugwood.org.

tions, the database is an excellent source for locating and identifying many organisms. Amateur scientists provide data about where invasive hotspots are and set out to tackle them. Recently, trained dogs, which can catch the scent of a small or inconspicuous seedling or insect, have even been recruited to sniff out unwanted species.

A good example of an "early catch" in Texas involves spotted knapweed (*Centaurea stoebe* ssp. *micranthos*). Texans had managed to avoid this highly invasive knapweed from Eastern Europe, which covers about seven million acres in 45 US states with its lavender, thistlelike flowers, until a visiting botanist came across a plant while hiking in Balcones Canyonlands National Wildlife Refuge in Travis County in June 2014.[3]

Having confirmed the plant's identity, the refuge's wildlife biologist at that time, Chuck Sexton, hastened to the site, which was an area vegetated with a commercial "native" wild harvest mix, originating in the central or northern Great Plains, where knapweed is already invasive (so buyer beware). Sexton located and destroyed three additional knapweeds and immediately notified the Texas Invasive Species Institute and TexasInvasives.org, which posted alerts. Canyonlands staff also scoured other parts of the refuge and found more plants. In 2015, personnel pulled up more plants and hope that the knapweed has been rooted out, although vigilance is ongoing.

So far, Texas has been lucky with spotted knapweed. Most species in this book have been here for too long and spread too far to be eradicated. They are here to stay. We may trap or shoot sparrows and starlings in specific sites, such as in orchards, in feedlots, and around purple martin colonies. We may wage a wider war on feral hogs and encourage hunters and land managers to shoot and trap as many as possible to slow down their increase and spread.

Traditional methods of animal control include hunting, trapping, poisoning, euthanasia, and sterilization; for plant control, burning, spraying, and mechanical removal. Best results often include combinations of methods. The precise one selected depends on the type of organism, habitat, topography, and climate (weather and season); type of threat to the environment; and its proximity to settled areas or native species—all idiosyncratic.

Biocontrol is the new, often most-preferred, management means and has a history dating back well over a century. Using natural enemies—whether predators, parasites, pathogens, or competitors—to reduce or control a pest species, biocontrol is often touted as a more effective long-term weapon. Applied carefully with ample research, using natural enemies has several distinct advantages. Unlike chemicals, the agent attacks only the pest in question. And unlike all other methods, once applied, biocontrol keeps on working. In the long run, this makes it cost effective, selective, and efficient.[4]

But biocontrol takes time to fully operate and may require considerable research backed by trials to predict its effectiveness and minimize the risks to native species. So biocontrol may be expensive over the short term and slow to implement. Additionally, no matter how much study goes into trials and tests, it is not always possible to anticipate how agents will behave in a new habitat or situation, especially decades—if not centuries—into the future. Biocontrol organisms are supposed to manage a pest, but as we have seen with grass carp, if they get hungry enough, they may consume native aquatic plants, too. Biocontrol experts work hard to release the right predator in just the right numbers, but in the end they are introducing a newer exotic to control an older one.

Most biocontrol agents are insects, moths, and beetles. Over the past 120 years, perhaps one in five of the 600 or so insects introduced into the United States has proven successful for biocontrol, notes biologist John R. Meyer. It has been a seesaw ride, he notes. Like the phorid fly that hovers over RIFAs, any successful agent must be host specific, survive in the release area, locate and synchronize its behavior and life cycle with those of its host, and reproduce in the wild. This is a tall order. Meyer adds that about one in three biocontrol agents has had limited success.[5]

We have noted that two weevils and a moth have reduced water hyacinth coverage in Gulf Coast states by as much as one-third. Researchers have found a fly that under optimal conditions shears through hydrilla beds. A cold-tolerant strain of weevil is having success with giant salvinia. Tamarisk beetles have done well in controlling saltcedar, though it may take a decade or more to kill off a thick stand. Over the last century, experts note, biocontrol agents have permanently or temporarily controlled at least 165 pests and weeds, and 170 biocontrol species are being sold globally to combat more than 100 serious pests, including aquatic plants.[6]

We are fortunate to have biocontrol agents for saltcedar, RIFAs, and the three freshwater weeds; however, there is currently no agent for Chinese tallow, which continues to spread largely unchecked. Nor are there comprehensive plans to deal with the tree except on a piecemeal or case-by-case basis. Feral hog research and commentary, which are detailed and voluminous, are now targeting a chemical toxicant that many experts hope is going to be the long-awaited breakthrough. The jury is still out (except in Australia, which is using the toxicant) while wild hogs are busily multiplying and spreading in Texas and in many other states. As mentioned previously, our attitudes and feelings toward a particular animal affect how we judge and manage it. We prefer feral cats to feral hogs but scratch our heads about dealing with their feline habits. Maybe if we feed wild cats well and keep tame ones indoors as much as possible, the situation will change so that we can enjoy cats and birds, lizards and bugs.

On the horizon, cutting-edge technology is being recruited to deal with invasive species. Since 2013, researchers have been investigating a manipulation whereby an invasive animal is able to pass on a fertility-reducing gene that may reduce overall numbers. Models suggest that this altered gene may prove aggressive, enabling a few individuals to spread it widely. For example, an altered weasel in New Zealand, where it is a serious menace, may spread the fertility-killing gene to other populations in New Zealand but also to other weasels in its native range after individuals have been released or have escaped. Pioneer Kevin Estvelt, head of Sculpting Evolution at MIT, worries that animals carrying the gene pass it to others that are not invasive, thereby making genetic alteration a risky proposition.[7]

Most of the time, especially on the local level, old-fashioned mechanical control is the simplest solution. Urban authorities and city neighborhoods are spending increasing amounts of time, energy, and money in cutting out or digging up nonnative species in parks, preserves, and sports fields. Community volunteers and enthusiasts identify; map the locations of nonnative species, especially plants; and then go out and remove them. This happens when agencies and the citizen-science programs they sponsor alert residents about local threats. A citizen-science group of volunteers in San Antonio, Texas, for instance, has been spearheading an effort to control invasive plants in its area. Volunteers meet for two hours on a weekday morning to chop the plants and clear them out. This form of neighborly outreach works well.

The Lady Bird Johnson Wildflower Center in Austin is a home for citizen volunteers who wish to invest time and energy in getting rid of invasive species. The center promotes an "Eradicator Calculator" (begun in 2009) on the TexasInvasives.org website that shows what is being done in regard to 149 listed plants and what results have accrued from rooting them out. Groups jazzily named Balcones Invaders, Cedar Ridge Choppers, Walnut Creek Wild, and others have directed more than 850 hours, for example, toward the removal of heavenly bamboo, or nandina, in a number of sites during the past six years. The Wildflower Center coordinates this Invaders of Texas program, training volunteers online and in person.[8]

One challenge in removing the same plant with other volunteers in a 21-acre city park in Austin in 2008 drew 3,800 volunteer hours and about $25,000 expended on digging up invasive plants. However, it is hard to identify and root out every seedling. Moreover, a plant may disappear, only to reappear after controls cease. This happened in the Austin facility. Local volunteers took out most nandina plants; however, many remained, adding up to 300–400 hours of extra work on the site.[9]

Successes against invasive species may often be short-lived. Enthusiasm tails off after volunteers turn elsewhere or a leader moves away. Although volunteers failed to stamp out the nandina, the example from Austin demonstrates how momentum is growing for protecting natural diversity, especially within urban areas. The TPWD has responded to and is promoting this enthusiasm.

John Davis, director of TPWD's Wildlife Diversity Program, notes that municipal officials, real estate developers, and concerned citizens are taking more and more interest in conservation on the local level. Davis and his group of urban biologists are working to address the so-called nature deficit disorder, meaning how we are increasingly alienated from contacts with the natural world. They are sponsoring efforts aimed at preserving, managing, and enhancing plant and animal diversity inside metropolitan areas. The goal is to promote individual involvement with native species and also to identify nonnative ones. Davis is proud of the state's model Master Naturalist Program, which has graduated at least 6,000 Texans under its biodiversity initiative. Part of this local-scale program includes the management and control of invasive species, notably pest plants.[10]

In this growth-pole corridor on Interstate I-35, City of Austin officials increasingly agree that having native plants and animals adds identity to communities and character to the state. Often these are in short supply. In some cases, as Davis notes, a feral pigeon may be the only contact inner-city children get with wild animals. But, he argues, the ubiquitous pigeon helps some youngsters become curious about other animals and plants they see, sufficient perhaps for them to take an interest in birds. Children observe city pigeons or city parrots and put food out or make a nest box for the schoolyard.

Instilling an interest in nature and natural habitats is a challenge, Davis admits, a bit "like bailing the *Titanic* with a thimble." But he is committed to opening up more urban areas for native species by protecting relict habitats in offbeat places and, where possible, through municipal buy-ins, restoring ones that have already been transformed, degraded, and neglected but not destroyed. He wants to return them to some semblance of a natural ecosystem. Davis is therefore committed to having people learn about a set of plants and animals, invasive or otherwise, that populate even the degraded habitats. "People decide the fate of their resources," he insists. Therefore, it is important to instill the political will and social values that contribute toward a better quality of daily and seasonal life for Texans.[11]

Opportunities for Rural Lands

Texas has more privately owned working land than any other state. Farms, ranches, and privately owned forests take up almost 83 percent of the state, consisting of 142 million acres. Farms and ranches are still being added, although the average size of properties has declined and now averages 520 acres. This means that there is ample scope for both old and new land-owners to tackle invasive species in rural areas. Opportunities are present through federal and state advice and financial assistance.

Federal agencies, most notably the US Fish and Wildlife Service (USFWS) and the USDA's Natural Resources Conservation Service, cooperate with state agencies in assisting landowners in the removal of invasive species. In Texas, the USFWS's Partners for Fish and Wildlife Program and TPWD's Wildlife and Inland Fisheries Division collaborate to fund and sustain a Landowner Incentive Program (LIP). With almost all the Lone Star State in private hands, which is much more than any other state, cooperation with property owners addresses a range of topics, including watersheds and riparian areas. In 2016, for example, the Texas LIP program had 18 active projects. They included efforts to conserve Guadalupe bass in the Pedernales, Llano, and Blanco Rivers and to remove nonnative chinaberry from bankside habitat. Through an alliance with Texas Tech and TPWD, LIP helped a landowner root out ornamental nonnative elephant ears on the South Llano River.

Farther north, a rancher in Donley County thanked both federal and state experts for helping him increase the number of wild turkeys and lesser prairie-chickens, together with other native plants and animals, on his 1,700-acre spread. He said, "The tremendous logistic and economic support provided by wildlife and agricultural organizations, Texas Parks and Wildlife, US Fish and Wildlife, NRCS [Natural Resources Conservation Service within the USDA], and Texas A&M Extension Service . . . improves the landscape for this and future generations. Instead of balance and trade-off, consider the bigger picture. Livestock, wildlife, conservation and habitat practices, and the organizations that assist us, are not competing but are complementary!"[12]

Improving overall conditions and increasing collaboration with biologists to sustain or improve natural biodiversity are ways that LIP helps local citizens. Removing invasive plants is important; so is supporting landowners for protecting endangered species. Addressing the endangered Houston toad, Benjamin Tuggle, a USFWS regional director, has noted that "private landowners are critical to the successful conservation of the Houston toad and providing them with tools such as the proposed Houston Toad Safe Harbor Agreement would give them peace of mind that their actions to conserve a part of Texas's natural heritage would not result in future restrictions on their lands."[13] In publishing the official Notice of Availability, the federal agency offered interested landowners in a nine-county area in east-central Texas a tool for brush management, woodland restoration, prescribed burning, wetland management, and invasive species control—all on behalf of the endangered toad. Importantly, to regulation-sensitive owners, the voluntary agreement assures that no new restrictions to those already agreed on will be added so that the property owner may return it to basic condition after fulfilling the agreement.

Tuggle's assurance responded to landowner pushback against what some state officers and many residents regard as government overreach. Helping landholders clear invasive vegetation or control exotic animals on behalf of an endangered species is one way of establishing trust among interested parties. Opening up habitat for the toad and other threatened plants and animals also develops opportunities for other native species and reveals problems that invasive species pose to these species.

The Environmental Quality Incentives Program (EQIP) within the USDA's Natural Resources Conservation Service is another voluntary program that provides financial and technical assistance about conservation practices to agricultural producers. These are people engaged in farming and forestry, who agree to draw up a Conservation Activity Plan (CAP) about a particular environmental concern. Cropland, rangeland, pastureland, nonindustrial private forestland, and additional farm- or ranchlands are included in EQIP.

An effort to bolster the numbers of monarch butterflies has been launched across 180,000 acres in 28 selected counties in Texas. The hope is to increase both food and breeding sites along the migratory track taken by the butterfly, whose population has crashed by more than 80 percent over the past 20 years. With financial and technical assistance provided by EQIP, landowners are helping restore early- and late-blooming nectar plants and also increasing native milkweeds on which the caterpillars feed. It appears that chemical pesticides have virtually eradicated native milkweed in the Midwest from a combined area the size of Texas in which monarch butterflies used to breed.[14]

The Urban Context

Cities and metropolitan areas are spreading across the face of the earth, both haphazardly and by design. More than 50 percent of the 7.3 billion people on the planet are now classified as urban residents; and that percentage will increase rapidly during the next several decades. Texas is no exception. Despite the state's vast acreage of privately held land, its population is overwhelmingly urban and will remain so. During the last 30 years, more than four million acres have been transformed and developed to accommodate the 500,000 new Texans who are making homes in the state every year. The cities and metropolitan areas that draw most of the newcomers already house 90 percent of existing Texans and offer a range of benefits to both people and plants and animals.

Another way of looking at invasive species is to understand what is happening in these humanly manufactured environments. They are massively transformed, historically occupied and managed sites, fashioned or rearranged by developers, planners, and entrepreneurs. We dam streams and rivers that flow through them; irrigate lawns and parks; construct buildings that resemble cliffs and canyons; and plant millions of foreign flowers, shrubs, and trees in suburban gardens, city streets, and satellite malls. We release or misplace our domestic pets and place feeders full of grains, seeds, and fruits to attract birds and butterflies (and also deer, raccoons, opossums, foxes, and rodents) into backyards.

Some native species have learned to exploit food and shelter and have begun breeding and loafing in urban settings. Raccoons, opossums, and native deer live in backyards and greenbelts. Common birds, such as wrens, titmice, and chickadees, nest in parks and gardens and visit feeders. Data suggest that more and more robins, mockingbirds, and grosbeaks are surviving colder months in northerly edges of their ranges because people are feeding them. As we have seen, white-winged doves and great-tailed grackles have also pushed northward in Texas in large part because cityscapes are good habitats for them. Even our largest metropolises are not inimical to wildlife. New York City, of all places, is now believed to have the largest urban population of peregrine falcons in the world. These native raptors, once decimated by DDT, nest on bridges and buildings and prey on city pigeons.[15]

Interestingly, coyotes live in almost every US and Canadian city. They are adept at concealing themselves by keeping away from people, usually but not always foraging at night, and setting up territories in fragments of wooded cover and rocky terrain in parks, neighborhoods, and greenbelts. Their territories are much smaller and often more fragmented than those of rural cousins due to readily available food, consisting of rodents, fruit, and deer and to a much smaller degree house pets. Juvenile coyotes survive longer in cities than in rural areas, where populations are often heavily persecuted. The American city's new wild canid is superbly adapted to urban life and successfully raises litters in well-trafficked parks, back lots, greenbelts, creek banks, and even culverts. Cats tend to avoid these areas because of coyotes, thereby enabling local songbirds to nest more successfully.

A new variant, the eastern coyote or coywolf, has been multiplying and spreading in recent decades. It is a cross between the western coyote and eastern gray wolf (with dog genes mixed in) and has colonized cities such as Boston, New York (one was captured in Central Park), Philadelphia, and Washington, D.C., extending into Virginia. This larger, stockier, longer-legged, medium-sized carnivore is reported to have originated in the Great Lakes region a century ago and headed east. Some experts believe they number in the millions. Coywolves hunt in both wooded and open terrain so blend into city scenery perfectly.[16]

For better or worse, the urban landscapes are also home for a host of nonnative organisms. Ornamental trees, shrubs, and flowers sprout in yards, gardens, window boxes, and increasingly on energy-saving green roofs, seeding and often spreading beyond their owners' properties. The plants add new shapes, structures, colors, and textures to bland neighborhoods, parking lots, and outlet malls. Some plants supply berries or seeds for hungry animals, such as newcomer parrots. As we have seen, bird aficionados released sparrows and starlings; pond enthusiasts spread water hyacinth; aquarium lovers brought in hydrilla and giant salvinia and freed the red lionfish. Owners lose or deliberately abandon mammals, reptiles, and fish and thereby encourage them to go wild. These, too, live in city neighborhoods, back lots, parks, rivers, and lakes.

Ecologists and others are discovering what types and how some nonnative species survive, adjust, and flourish in urban environments while others dwindle and disappear. Experts have compiled lists of native and nonnative species that adapt to downtown areas, commercial districts, shopping malls, old and new residential neighborhoods, industrial parks, golf courses, transportation routes, water-treatment plants, and even landfills. Each urban site is a microhabitat for both native and nonnative organisms. Urban ecologists are also determining how some species, whether native or not, associate or compete with one another and how human residents respond to their presence. Positive or negative attitudes arise out of personal experiences and community expectations about what native and nonnative species do. Educational outreach is informing residents about the benefits and costs of having all kinds of plants and animals inside neighborhoods. Residents are becoming more aware about how landscapers or garden centers and pet stores contribute to the mix of native and nonnative plants.

Recent research on birds in metropolitan areas, for example, shows that a relatively small group of highly adaptable species, such as pigeons, sparrows, and starlings, occupy downtowns. Then, as one heads away from the usual noise, grime, and congestion, additional species, native and nonnative, show up. And as a city pushes rapidly into the countryside, the character of the indigenous birdlife continues to show up but grows more splintered and localized.[17]

Some researchers categorize the urban birds as either exploiters or adapters. Exploiters are bird generalists that quickly take up residence and spread inside cities. They comingle with inner-city exotics, such as sparrows, pigeons, and starlings, and are often joined by native specialists, such as nectar-feeding birds (hummingbirds) and birds of prey (owls and hawks) that find abundant food within city limits. Exploiter species prove to be well-recognized urban residents. Great-tailed grackles and blue jays perform these tasks in many Texas cities.

Adapter species tend to move into a city more slowly, occupying habitats that resemble the same ones to which they are adapted. Rather than disappear, they take up residency in new landforms, such as urban lakes. Pied-billed grebes, summering green herons, and yellow-crowned night-herons, for example, are living inside many Texas cities, just as American coots and double-crested cormorants and several duck species commonly winter in the same watery places.

Studies suggest we should pay close attention to adaptive species and the places they inhabit because they highlight the basic character of a city's overall birdlife. And because both native and nonnative species coexist, mingle, and have different impacts on the environment, bothering about their origins becomes irrelevant.[18]

There are surprises about this re-sorting that are not limited to birds. At least one-third of California's 236 butterflies have begun feeding on non-native plants, of which more than 1,000 species are now naturalized in the Golden State. This fraction includes at least nine butterfly species that have prospered in urban places. The insects visit nonnative plants, notably ornamental flowers and shrubs, and in several cases have reportedly come to depend on them.[19]

Contemporary examples of native wildlife in the city continue to astonish us. Recent work in the United Kingdom describes a mosaic of habitats in urban settings (now covering 7 percent of that nation's land area) that host more wild bee species than rural areas do. While this fact is an indictment of modern industrialized monocrop-style agriculture and insecticide sprays, it is also a celebration of how metropolitan environments may be islands for biodiversity and abundance. The density of red foxes in England reaches 14 per square mile in the cities, compared with only 1 fox per square mile in rural areas. A fox, nicknamed "Romeo," decided to den on the seventy-second floor of an unfinished skyscraper in London and survived on food scraps left by construction workers. Current cityscapes are cleaner and less polluted than equivalents or counterparts 50 or 60 years ago. Human residents are being offered unheard-of opportunities for learning about and observing wildlife that is returning. Berlin, Germany, has more fox dens than in the surrounding forest and has become a home for white-tailed eagles, not seen for nearly a century. Eurasian beavers, native to Germany and its neighbors but hunted almost to extinction a century ago (a few remained along the River Elbe), have shown a remarkable resurgence because of transplants and protection; they also are making Berlin home. Stockholm, Sweden, has a healthy population, and some are now breeding in Britain after being trapped and hunted out 500 years ago.[20]

All over the world, urban lands are acting as natural refuges for animals, once residents accept them. South Africa's Cape Town is home to 11 troops of native chacma baboons. Three dozen leopards forage among 21 million people in Mumbai, India. Ten times more brown bears (grizzly bear equivalents) inhabit 22 mostly industrialized European countries than exist in the lower 48 US states. Until 20 years ago, no wolves existed in Germany. A healthy pack now lives within 30 miles of Germany's second-largest city, Hamburg, and pups have been sighted even closer to downtown Berlin. "Cities are possibly the most exciting, most surprising, and least understood ecosystems on the planet. They are places where much of what we think we know about the natural world doesn't apply . . . places that may even be changing the animals within it, just as the shift from country to city changed us," says Tristan Donovan, author of a book on city wildlife.[21] Animaltourism.com backs up his opinion. The website lists sites in different countries where various species may be observed. Many are within or close to urban areas and show that many wild animals are more approachable than ever before so change the ways we think about them.[22]

Virtues of Watch and Wait

This urban shuffling and re-sorting by no means suggests we should throw in the towel about invasive species. Disturbed environments are multiplying and spreading across the face of the earth. This is all the more reason for us to pay attention to which plants and animals inhabit city spaces. State and municipal outreach, neighborhood programs, and citizen-science initiatives should continue to invite residents to identify, survey, assess, and then carefully target noxious plants and animals, usually invasive nonnatives, for control and even extirpation. However, in some contexts a wait-and-see approach provides more nuance than the overplayed meme that categorizes native as good and alien as bad.

A watch-and-wait policy is based on acceptance that metropolitan areas are highly disturbed environments occupied by both native and alien plants and animals. Singling out an invasive species like Chinese tallow or the house sparrow is often a waste of time and money. Rather, it may be better to observe and document what the alien tree or bird is doing to local biodiversity and then, if deemed problematic, rely on a specific method to control or get rid of it on a case-by-case basis. Before waging war on coyotes because people have traditionally disliked them, it is better to wait and see which animals create problems for homeowners and their pets and then target the individuals.

Geographer Kevin Anderson, who heads Austin Water Utility's Center for Environmental Research, notes there are always problems to be faced and dealt with in ecosystems, whether rural or urban, natural or humanly created. Anderson believes that reliance on a native-species approach is too limiting; too much has changed, too much has been disturbed, he says, not to consider and sometimes acknowledge benefits of having nonnative species. Moreover, we do not know what climate change will do to urban wildlife or how far we can think ahead. Few people anticipated the sudden spread of the white-winged dove in Texas or documented an earlier one by the great-tailed grackle, or of native mesquite, or the (once mainly Mexican) nine-banded armadillo. "Dynamism marks systems, not permanence," Anderson concludes.[23] In this view, we should avoid condemning nonnative organisms simply because they are newcomers to a landscape. They have roles to play, so it pays to watch and monitor what these are instead of assuming they are always negative.

Watchfulness may also help prioritize offenders. Austin lists 24 invasive plants—ferns, grasses, shrubs, trees, and aquatic species—but officials are reluctant to prioritize them due to a lack of uniformly robust data. However, if recent statements are true, waging war is not only costly but becomes a series of never-ending skirmishes. For instance, the city's Parks and Recreation Department has taken action against invasive species on about 7 percent of the 15,000 acres for which it is responsible. Its Watershed Protection Department has conducted operations on 12 percent of the prioritized waterways. Treatment costs around $1,000 per waterway mile. Another way of looking at these figures is to say a watch-and-wait mind-set helps target controls more efficiently.[24]

In California, migrating monarch butterflies are predominantly wintering in Australian eucalyptus trees in more than 200 coastal sites from Mendocino County south to San Diego County. Imported from Australia, the eucalyptus trees were obviously unavailable to monarchs before enthusiasts, notably Ellwood Cooper, planted them after 1850. The monarch butterfly has shifted to widely dispersed blue gums after logging took out native pines and cypresses. Currently, eucalyptus furnishes critical shelter during cooler months when the insects are less active. This situation recalls the nonnative saltcedar in which the endangered southwestern willow flycatcher builds its nest rather than in native willows and cottonwoods. Cutting down these exotic trees without regard for how they function for butterflies or flycatchers may hurt a highly regarded native species. In fact, monarch butterfly populations are in trouble east of the Rocky Mountains, and their coniferous forest roosts in Mexico's state of Michoacán are being splintered. Thus, a comprehensive and contextualized assessment of how a plant fits into a modern urban landscape, even if exotic, deserves close inspection, especially after native habitat has been lost or totally degraded.[25]

Interestingly, between 2005 and 2015, the numbers of nesting house sparrows have trended *downward* in Texas, although they have held more or less steady for European starlings (which have increased in the Edwards Plateau and South Texas Brush Country). A downward trend across the nation as a whole for the sparrow and a smaller drop-off for the starling are welcome news. Whether this decline in sparrows is somehow linked to a larger drop-off in the United Kingdom—where their numbers in cities have fallen by 60 percent in the past three decades, placing the bird on Britain's Red List of endangered species—calls for further study.[26]

Being watchful pushes the need for understanding. Controlling species is more than plug and chug—a mechanical or automated response to a

Monk parakeet or quaker parrot. Photo by Bernard Dupont (CC-SA 2.0).

problem without creating or risking a new one. The Great Lakes basin's 187 nonnative plants and animals, some of which we released to deal with earlier releases, are a case in point. We need to study any new flora and fauna carefully, to see not merely how they function and interact with native species but also how changing conditions, such as drought or wildfire, may also change what they do. Calling on the experiences of dedicated fellow citizens, some of whom commit years of fieldwork and research to a particular animal or plant in a specific location, is invaluable. Listening to resident opinions breaks down customary barriers between professional experts and experienced laypeople, who may have long expertise about a set of conditions and be prepared to share it.

The Monk Parakeet

We began this book with the fluttering antics and querulous calls of the monk parakeet, and it is also a fitting place to end. The green-backed, gray-breasted parrots are, of course, not part of Texas' indigenous and famously large avifauna. They are exotics from South America, notably Argentina, Uruguay, and neighboring parts of Brazil, that reportedly escaped in Austin from an RV Park along Barton Springs Road close to Zilker Park in the early 1970s and nested in ballpark light fixtures along Town Lake, as it was then called. Elsewhere, pet fanciers released other birds so that the feisty

Monk parakeet inspecting its enormous nest (approximately five feet long) in Austin, Texas. Photo by Matt Turner.

parrots inhabit 15 US states as well as several European nations. They are an urban species, and in Texas they live in Dallas, Fort Worth, and Houston as well as in Austin, where at last count there are least 100 nest sites, some of which may have 30 breeding pairs. There is hardly a spot in central Austin where you do not hear their telltale squawks. Voluminous stick nests may approach the size of a small car and have at least on one occasion caused a fire on a transmission pole, knocking out power to 1,000 customers.[27]

Since they are not native to Texas, they already meet half of the requirement for invasive status. They clearly are doing more than merely surviving. Now that they have begun to occupy transmission towers into which they weave twig nests, they seem to be thriving, and their numbers are increasing. Are they causing harm? Purists might say so. The monk parakeet does not belong here and takes up resources, feeding on a range of plants, including berries, grass seeds, and fruits that some other bird or mammal could be using. Who knows what animals it might have displaced within its many nest areas. This is surely some sort of environmental harm, even if poorly quantified. And the power outage is economic harm. Several states have already made it illegal to sell or own the parakeets for fear they will become pests, and British authorities have systematically shot, trapped, and snared 100 or so birds located mostly in London.

The lessons from this book suggest that in the absence of exploding numbers and explicit and repeated harm, such an aggressive stance may be unwarranted. Standing back a little and adopting a nuanced approach in regard to nonnativeness is not necessarily a mistake. Seeking to describe and understand the ecologies of urban places, the ones that most of us reside in, involves noticing the many living things that surround us. Learning how a nonnative species, such as the parakeet, a fire ant, or aquatic weed, functions in its source region and in its new home is essential for formulating and prioritizing management decisions. How a species lives and moves within its new range may differ from how it lives within its native range or even in other places where it has been introduced. Therefore, because a species is nonnative does not immediately preclude us from accepting it. Nor should we necessarily demand its eradication, as some people are currently doing about eucalyptus in California.

The information we may gain from discovering the functions or environmental services a plant or animal performs may be a pleasant surprise or benefit. In South America, the monk parakeet is a crop thief. In the United States and Puerto Rico, the bird is an urban species. The parrot has not earned enmity among farmers. Nobody knew that it would nest on metal light poles in stadia and then occupy transmission towers that are blossoming across the United States. There are financial costs associated with this habit, but so far they are not high. Some enthusiasts are weaning some birds away from electrical equipment by providing alternative nest sites.

There are also cultural issues at stake. City folk may be comforted that amid the snarling traffic, sidewalk weeds, starlings, and pigeons, an unusual bird, which we tend to erroneously associate with Amazonian tropical jungles, is a fellow urbanite. There are benefits from considering parrots as attractive, interesting, and gregariously inspiring companions in our daily lives. In many ways we consider them closer to pets than wild birds, as any visit to a pet shop demonstrates. Local residents appreciate different parrot species because they add color and verve to a neighborhood, such as in McAllen, Brownsville, and Edinburg in the Rio Grande Valley, where at least seven species live, or on Telegraph Hill in San Francisco, with its famous population of red-masked parakeets. There are additional parrot watchers elsewhere, including Los Angeles and Miami—urban islands with thriving populations of interesting, chattering birds, some of which are endangered in their native ranges. Almost 100 species of parrots exist in cities around the world. Through this synurbanization—a process by which animals are adapting to urban life—bird lovers want to conserve them. Thus, there is no need to condemn them for being foreigners.[28]

This position does not mean nonnative parrots are somehow "improving" nature, as some of our predecessors argued. We have referred to the sins from misinformed decisions by introduction zealots. It means that introduced birds are breeding in the United States, adding character and

The now extinct Carolina parakeet was once abundant in Texas and the eastern half of the United States. Illustration by John James Audubon, *Birds of America* (1833), plate 26.

vibrancy to streets and neighborhoods. It also means that as city parrots grow more common, they blur the line between what is native and not native. Their presence may also check hasty judgments and vehemence toward the "invasiveness" of nonnative species, because in a relative sense not many of them fall into this category.

Sadly and ironically, we have already destroyed the only parrot endemic to what is now the United States, the once-abundant and widespread Carolina parakeet. With a reddish-orange face, bright yellow head, and green to blue-green plumage, the foot-long parrot once flew across mature bottomland forests and cypress swamps in the eastern half of the United States, from New York south to the Gulf states and as far west as Colorado. A stunning illustration by Audubon depicts a cluster of these parakeets that he had shot feasting on cockleburs. Explorers and colonists reported raucous groups of 200 to 300 birds well into the 1800s, but the felling of the eastern deciduous forests and trouble with farm crops led to the birds becoming scarce. In East Texas, old-timers remembered them as plentiful before the Civil War; but farmers shot them and set up scarecrows to keep them away from their corn patches. Famous birdman Henry Philemon Attwater reportedly came across a single Carolina parakeet near Beaumont after 1910. The last credible sightings were in Florida in the 1920s, and after that the Carolina parakeet was generally considered extinct. Not a single comprehensive biological study of the bird was published before it disappeared.[29]

The squawks and antics of the monk parakeet are a reminder, not only of the poignant loss of a native species but also of the resilience and mutability of the natural world in a modern, urban context. This is not to ignore or belittle the fact that more and more alien species are moving around the world than ever. So cooperating on international protocols and procedures that address the impacts of the "100 worst ones" to forestall their arrival and spread is vital. Knowing how they move and which commercial or other pathways they take reduces the potential damage they are likely to do. In this way we can target scarce resources on a problem organism while its numbers are small or localized, while accepting or even embracing other nonnative species as members of the communities in which we also live and work.

APPENDIX 1
Who Deals With Invasive Species?

International Controls

Achim Steiner, a United Nations (UN) official, estimates that globally pests cost upward of $1.4 trillion annually and states that far too many nations have underestimated or been casual about threats invasive species pose. Three international conventions dealing with mammals and birds include invasive species. One of them, the Convention for Migratory Species of Wild Animals (1979), aims to stop invasive species from affecting wild animal populations, notably birds.[1]

The second one, the UN Convention on the Law of the Sea (1994), addresses nonnative and invasive species in the marine environment. This convention requires its 168 signatories and the European Union to keep all coastal waters and seas free of invasive organisms. The United States has signed the agreement but not ratified the convention.[2]

Third, the Convention on Biological Diversity (CBD), launched in 1992, is now the legally binding cornerstone for alerting 196 parties about protecting, managing, and sustaining the wealth of the planet's genetic resources. The CBD is concerned that alien species are undermining this natural health. The United States has signed but not ratified the convention; it enjoys "observer" status.[3]

Trade-related bodies, such as the World Trade Organization and the International Maritime Organization, are becoming more proactive about keeping alien and invasive species out of commerce and transportation. The International Civil Aviation Organization, a UN Specialized Agency, for example, aims at stopping nonnative species hitching rides on aircraft by establishing quarantines and disinfection procedures (the brown tree snake's arrival on Guam by aircraft is often mentioned). The UN's Food and Agriculture Organization has a program to identify and combat plant and animal pests in forests and woodlands. And Botanical Gardens Conservation International identifies similar threats in urban parks and reserves and trains people to conserve and restore healthy plant communities. Keeping as many habitats, terrestrial, riparian, and marine, as clear as possible is increasingly a priority among these international agencies. They acknowledge that the public appreciates the economic, health, and environmental issues at play.

Several organizations release lists of invasive plants and animals. One of them is the Invasive Species Specialist Group, established in 1994. The group is part of the Species Survival Commission of the International Union for the Conservation of Nature and Natural Resources (IUCN). A pathbreaking semiofficial agency founded in 1948, the IUCN has produced a well-publicized "100 of the World's Worst Invasive Alien Species," of which a dozen live in Texas.[4]

The United States

Federal and state agencies, together with nongovernmental organizations, are major players in establishing policies and programs aimed at controlling invasive species in the United States. Six departments, including Homeland Security and the US State Department, have jurisdiction over alien species. Agencies, offices, institutes, field stations, committees, and programs within cabinet-level departments assemble, collate, and archive data about invasive species and the problems they pose. They also issue regulations about preventing entry and spread and support programs to get rid of nonnative species.

There are specific collaborative programs controlling invasive species. For example, the Great Lakes Commission (an interstate agency set up in 1955) works with the US Environmental Protection Agency (EPA) and US Great Lake states and Canadian provinces in a binational effort to manage approximately 185 nonnative water-dependent organisms that have gained access to the St. Lawrence Seaway and the Great Lakes during the last 200 years.

Housed at the US Geological Survey (USGS) in Fort Collins, Colorado, the National Institute of Invasive Species Science (NIISS) makes data about invasive species accessible to both land managers and the public. The NIISS is a hub for science collaboration, coordination, and integration across agencies and disciplines.[5]

NASA's Global Organism Detection and Monitoring system within NIISS pinpoints records of invasive species throughout, but not exclusively in, the United States. The institute's website supplies information about the name, location, program, and map of an organism. The aim is to streamline information among lead agencies so that scientists and other committed parties can track invasive species and establish points of collaboration.

There are 43 congressional laws and a larger body of amendments to existing laws that address the entry and spread of unwanted and invasive organisms. Several legal instruments name species, such as the tamarisk or saltcedar, Russian olive, brown tree snake, Asian carp, sea lamprey, nutria, and European starling. For example, the US Fish and Wildlife Service (USFWS), US Bureau of Reclamation, USGS, EPA, and US Army Corps of Engineers refer to statutes to try to limit the spread of the zebra mussel by people.[6] Others are more general. The oldest law is the 1900 Lacey Act, which prohibits the importation and interstate transport of noxious species and, as of 2008, includes a variety of plants and wood products taken illegally from any US state or foreign nation.

In 2012, the Lacey Act, revised and amended, listed more than 230 animals considered harmful to public health, agriculture, forestry, or native wildlife. The law instructs the secretary of the interior to regulate the importation and transport of fish, including the walking catfish and several carp, mollusks, and crustaceans, totaling 136 species; 92 mammals, including the originally listed mongoose, flying foxes, and fruit bats; and 6 birds, including the house sparrow and European starling. Five reptiles, headed by the brown tree snake, which has done so much damage to birds on Guam, and the Burmese python in the Everglades round off the current list.[7]

The Lacey Act names the USFWS as a key responder. In order to publish in the *Federal Register*, the USFWS gives notice about an impending inquiry. Then after data review and assessment, it invites public comment about a species before it rules on a listing. This stepwise approach is a guide to adding or dropping a species from its injurious species list. It also helps the USFWS Wildlife Inspection Program update what can legally pass through the nation's 38 major ports of entry.[8]

The Non-Indigenous Aquatic Nuisance Protection Act of 1990 (renamed the National Invasive Species Act in 1996) directs specific programs toward aquatic weeds and animals, has a Nuisance Species Task Force to manage them, and addresses ballast water releases. The law resulted from zebra mussels infesting the Great Lakes.

The importance of being more vigilant and responsible grew after President Bill Clinton signed Executive Order 13112 in February 1999, which drew attention to the impacts of alien organisms on the agricultural and environmental fabric of the nation. It established the National Invasive Species Council (consisting of secretaries and other key personnel of 12 federal agencies) and authorized the council to publish a National Invasive Species Management Plan that provides leadership about invasive species. The plan examines existing and prospective approaches for dealing with the unwanted organisms and identifies the agencies best suited for preventing the introduction and spread of them, including pathways used. The plan for 2016–18 prioritizes actions needed to control invasive species and restore areas affected by them. A new Executive Order 13751, signed by President Barack Obama in December 2016, amended the original order to expand membership of the National Invasive Species Council and added considerations about climate change, human health, and the development of new technologies for dealing with invasive species.[9]

Texas

Administrative actions among state agencies help streamline efforts to control invasive plants and animals. For example, the Texas commissioner of agriculture oversees programs against alien crop pests, and his agency participates in the national Cooperative Agriculture Pest Survey (CAPS) linking Texas to the USDA. CAPS looks for invasive insects and pathogens, such as sweet orange scab, the Old World bollworm, channeled apple snail, plum pox virus, rice pathogens, nematodes, and other pests.[10]

The Texas Department of Agriculture (TDA) administers more than 50 laws concerning agriculture, so controlling plant pests and crop diseases is a major responsibility. Under the Texas Agriculture Code titled "Noxious and Invasive Plants," for instance, TDA lists 28 harmful plants it has targeted for control and eradication, including Chinese tallow, saltcedar, and the water hyacinth. The list identifies 26 noxious plants (4 of which are also classified as invasive), which can be native or introduced and likely to do economic and/ or ecological damage to farmers, ranchers, and the markets they supply. Two recent additions, chinaberry and Japanese climbing fern, are labeled simply invasive because of their general absence from cropping areas. TDA evaluates additions to the list submitted by agencies or members of the public.[11]

The Texas Invasive Species Coordinating Committee (TISCC) facilitates efforts to prevent and manage invasive species. Established by House Bill 865 (May 2009) by the 81st Legislature, TISCC members include representatives of six state agencies, including the AgriLife Extension Service, Forest Service, and the Water Development Board. Members meet in Austin every three or four months. The committee stands behind the "Noxious and Invasive Plant" list drawn up by the TDA and the "Invasive, Prohibited and Exotic Species" list drawn up by the Texas Parks and Wildlife Department. Species overlap exists between the two official lists.[12]

In the nonofficial sector, the Texasinvasives.org website proclaims dramatically, "Hello Invasives. Goodbye Texas," implying that pest species are eroding the state's identity as well as its economy. The invasive species group partners with federal and state agencies, universities, and other stakeholders to manage nonnative invasive plants and pests. It houses a database of invasive species that lists and maps about 100 plants, 20 animals, 45 insects, and 4 pathogens and keeps users updated about their status and ongoing problems. The organization recruits volunteers to check outbreaks or reports of noxious species, encourages best-practice activities outdoors that minimize contacts with them, and e-mails subscribers about news and events. The group also organizes conferences hosted by one of its listed partners, such as the Texas Invasive Plant and Pest Council (TIPPC) established in 2008. The TIPPC is also a forum for information exchange and research support.[13]

Local groups are also active. The Austin Invasive Species Coalition hosts a website for capital-area events and promotes plant biodiversity through the elimination and control of invasive species. It posts links to other groups that are also battling these species. In 2010 the Austin City Council passed a resolution to develop an Invasive Species Management Plan aimed at controlling aquatic, riparian, and terrestrial species, initially plants. The council aims to coordinate volunteer, academic, and municipal authorities under an umbrella that covers issues of local concern.

The North Texas Water Garden Society, based in Dallas, lists prohibited aquatics and suggests "the spread of exotic plants is an extreme threat to the native aquatic environment and a potentially dangerous situation." The University of Texas at Austin Lady Bird Johnson Wildflower Center tracks alien species through its "Texas Invasives" database.

APPENDIX 2
Invasive Species in the Americas

1493 Columbus sails with cattle, chickens, sheep, goats, and pigs on his second voyage to the New World.

1519 Hernán Cortés puts horses ashore on mainland Mexico.

1519 Cortés takes the New World turkey back to Spain, where it becomes well established. Introduced into the United Kingdom in 1524 and bred successfully, the domesticated bird is transported back to North America, beginning in Jamestown, Virginia, in 1607.

1538 Viceroy Antonio de Mendoza and Cortés hold a banquet on Mexico's mainland at which dignitaries feast on goat, ham, quail, chicken, mutton, beef, cabbage, chickpeas, and Spanish wines.

1560 Olive seedlings are imported into Lima, Peru, from Spain.

1602 Wheat and oats are first planted on islands off the coast of Massachusetts.

1609 European black rats infest Jamestown, Virginia. The house mouse follows.

1622 European honeybee is reportedly carried from England into Jamestown, Virginia, and by 1638 was also introduced into Massachusetts, becoming widespread in New England after about 1675. Thomas Jefferson noted that Native Americans call it "the white man's fly."[1] By 1800, bees have swarmed into Indiana, Michigan, Illinois, Iowa, Missouri, and Texas, where the bee is the official state pollinator (2015), reportedly carried in by Spanish missionaries.

1634 Sorrel and yarrow, Old World medicinal plants (yarrow has a North American genotype), escape from cultivation in New England.

1650–1750 Red fox is imported into the United States from England on several occasions.

1700 Feral hogs are described as "swarming" in Virginia. By 1834, some 60,000 hogs roam the woods around Nacogdoches, Texas, outnumbering people seven to one.

About 1700 Some 48 million wild cattle are grazing the Pampas in Argentina. Today, only one-quarter of Pampas plants are native to Argentina; Old World grasses and clovers dominate the eastern, moister section.

About 1723 A French naval officer carries coffee seedlings from France (Jardin des Plantes in Paris) to Martinique, which becomes the source for growing coffee plants in the Americas.

1772 Benjamin Franklin sends seeds of the Chinese tallow to Savannah, Georgia, for planting as a source of oil. By 1824, the tree was naturalized in coastal Georgia and South Carolina but spread aggressively only after the USDA's Bureau of Plant Industry and others promoted tallow for oils and soap after 1900. The tree was likely in Houston, Texas, by 1910. It was cultivated as a shade tree in the Rio Grande Valley by 1925; the Texas Highway Department also planted it for shade along highways starting in the 1940s.

1785 Spanish introduce date palms into Mexico and Baja California.

1790 French botanist André Michaux establishes a garden near Charleston, South Carolina, where he plants both crape myrtle and chinaberry, both native to East Asia; the latter proves to be invasive.

Early 1800s St. John's wort, a perennial medicinal herb, is introduced into New England and spreads; currently nine US states consider it invasive.

Early 1800s Spanish cane or giant reed is planted in California. It came from Mediterranean Europe, where it is cultivated for erosion control and roof thatching. Cane is invasive in seven US states, including Texas. In 2007 a European wasp was introduced to control cane stands in California and in the Rio Grande Valley.

1818 A species of saltcedar is growing in the Botanic Gardens at Cambridge, Massachusetts. By 1823, nursery catalogs on the East Coast are promoting saltcedars, native to Africa, the Middle East, and Asia, and by 1850 West Coast catalogs follow suit. Trees are growing in Galveston, Texas, in 1877, and during the next decade the US Army Corps of Engineers spreads them on barrier islands.

1832 Charles Darwin describes vast tracts of Uruguay dominated by Old World artichokes, a thistle.

1832 Brazilian peppertree is available in ornamental plant catalogs in New York City; imported to the Miami area in Florida in late 1890s; and is growing in Brownsville, Texas, 50 years later. The peppertree is a refuge for alien ants and a root weevil.

1839 Commissioner of the Patent Office of the State Department (the progenitor of USDA in 1862) secures congressional funds for the collection of the "seeds of new and rare varieties of plants." The commissioner of patents reports the introduction of Baden corn and other wheat varieties. By 1880, sorghum, Kafir corn, various wheats, sugarcanes, flowers, and vegetables had been introduced and distributed. The overall cost of seed and plant and introductions under US government direction was estimated at $4.5 million from 1852 to 1905 with returns valued in excess of $100 million in 1905 alone.

1850–53 House sparrows are imported and released in New York City. In 1867 T. H. Brown releases some birds imported from Liverpool, England, near his house in Galveston, Texas.

1851 Commander of the East India Squadron of the US Navy sends back sugarcane cuttings that are distributed to planters in Louisiana.

1857 A special agent is sent to Japan to collect tea seeds and study tea plants for trials in the United States (failed after 30 years of tests).

1864? Sorghum is introduced to United States from China and France under government sponsorship.

1868 Gypsy moth from Europe is released into Medford, Massachusetts, as a disease-resistant silk-spinning caterpillar hybrid. In 1981, almost 100 years after the first outbreak, the moth defoliates 13 million acres of the US Northeast woodlands, mostly hardwoods; strips 320,000 acres in New Jersey in 2007; and continues to kill and damage trees.

1871 Striped bass, an anadromous Atlantic Coast fish, is introduced to the US West Coast and forms a sport fishery. By 2000, the bass has been introduced into at least 450 inland reservoirs in 36 states, including Texas, during the previous 30 years.

1876 Kudzu, or Japanese arrowroot, a leguminous vine, is touted in the Japanese pavilion at the Continental Exposition in Philadelphia as a forage crop, shade giver, and ornamental plant. Now characterized as a noxious weed in the United States, this member of the pea family covers about 7.4 million acres of the US Southeast and is spreading over 150,000 acres annually. It serves as a host for Asian soybean rust and is likely to benefit from extra atmospheric carbon, thereby spreading the invasive rust.

1882 US Bureau of Fisheries introduces the European or common carp into Texas and promotes its propagation.

1884 Water hyacinth, native to Brazil, makes its alleged first US appearance at the World's Fair in New Orleans as a souvenir distributed by Japanese officials. By 1890 the hyacinth had invaded Florida's waterways, and by 1902 Congress had authorized funds for its removal from navigable waters in Texas. The floating ornamental weed is naturalized in all southeastern coastal states.

1886 European hares are introduced into Jobstown, New Jersey.

1890s Cheat grass, an annual bunch grass from southwestern Asia, enters the United States via Europe and is soon noted in New York and Pennsylvania and reported in British Columbia, Canada. By 1930 the grass covers 60 million acres in the Great Basin and is dominating arid grasslands in Oregon, Washington, Idaho, and Nevada. It is also widespread in Canada and northern Mexico.

1890 About 60 common, or European, starlings are freed in Central Park, New York City, under the auspices of the American Acclimatization Society and followed by 40 more in 1891. Releases in Portland, Oregon, at about the same time do not appear to become well established. The first starling shows up in Cove, Chambers County, Texas, in 1925.

1890 Thirteen wild boars from Germany are released in New Hampshire. Several boars are set free on a ranch in Calhoun County, Texas, in 1939, and with later releases on the Edwards Plateau they interbreed with feral hogs.

1898 USDA establishes a Section of Seed and Plant Introduction and Distribution (later housed in its Bureau of Plant Industry). Over the decades the bureau imports many cold-hardy and drought-tolerant plants useful to agriculture, including Kyushu rice from Japan, as well as exotic ornamentals, some of which have become invasive, such as honeysuckles, privet (*Ligustrum* spp.), and the Brazilian peppertree.

1899 Nutria, or coypu, from South America are introduced into Southern California and released in Louisiana in 1933 (and at other times during the 1930s) from where they spread.

1900 An epidemic of bubonic plague breaks out in San Francisco, killing 113 people. The disease infects ground squirrels and is carried by the fleas of 34 rodent species in the US West.

1900 USDA personnel visit Algeria to study date growing, having visited Baghdad two years earlier.

1904 Chestnut blight, a fungus from Asia, is found in the Bronx Zoo, New York City, and within 40 years eliminates 3.5 billion chestnut trees in North America. A backcross system using resistant Japanese and Chinese trees provides partial immunity.

1904 USDA, Bureau of Plant Industry, sends more than 10,000 bushels of cottonseed to Texas to help fight the boll weevil. It is also testing varieties of alfalfa (obtained from Turkestan in 1897) for nationwide use, as well as plants used for camphor and morphine, vanilla beans, mangoes, and cassava for Florida. The bureau operates testing stations in Arlington, Virginia; Tempe and Yuma, Arizona; and Mecca and Chico, California, with secondary sites in San Antonio and Chillicothe, Texas.

1910 Congress considers importing African hippos into Gulf Coast swamplands to check water hyacinth and serve as meat (the bill fails to pass).

1912 Date palms purchased by private individuals in Iraq and destined for California's Coachella Valley are unloaded in Galveston, Texas.

1918 The black imported fire ant from South America appears in Mobile, Alabama.

1924 Nilgai antelope, native to India, are shipped from the San Diego Zoo to the King Ranch in Texas. Exotic game ranching in Texas expands in the drought of the 1950s. By the 1990s, exotic game numbers 164,000 (among 67 species) inhabiting 2.3 million acres on 486 ranches in 137 Texas counties.

1928 Dutch elm disease, an Asian fungus spread by beetles, enters North America on logs sent from the Netherlands (it had first entered Europe in 1910) and spreads. By 1990, it has killed about 75 percent of American elms. In 1967, a second more serious strain originating in North America arrives in Britain and kills off more than 25 million trees. France also loses more than 90 percent of its elms. A major outbreak occurs in New Zealand in 2013.

1932 Red swamp crawfish, or Louisiana crawfish, native to fresh waters of US Gulf States and Mississippi Valley, is introduced to California and has spread to 15 more states, as well as to Europe (it is now a pest in Spain), Africa, East Asia, Central and South America.

1933–41 The red imported fire ant, native to Peru, Brazil, and Argentina, appears in Mobile, Alabama. Future Harvard entomologist E. O. Wilson, still a high school student, studies the new insect. The ant hit Texas by 1953, as well as nine other states.

1935 Cane toad, native to Central and South America, is shipped from Hawaii to Australia to control the cane beetle. It proves invasive.

1938 Viral hemorrhagic septicemia, which kills more than 50 freshwater and saltwater fish species, is noted in Europe, eventually spreading to 38 nations. It spread to Pacific Coast salmon in the United States (Washington State in 1988) and other fish in the Great Lakes (after 2003).

1951 Hemlock woolly adelgid, an aphidlike insect from East Asia, begins to kill two hemlock species in New England. By 2007, it occurs in 50 percent of the eastern hemlock's range (16 states). Insects from Japan, China, and the western United States are being introduced to kill off the sap-sucking pest.

1951–52 Hydrilla, native to Asia and Australia, is imported by mistake to St. Louis, Missouri, and soon after traded to Tampa, Florida, where it is marketed in the aquarium trade as Indian star-vine. By 1960, waterways in the Miami area are choked as the plant spreads quickly via boat trailers, turning up in Texas in 1975.

1953 Of the 89 common weeds in New York State, 35 are from Europe.

1957 Some 26 swarms of the Africanized honeybee hybrid escape from confinement in Rio Claro, São Paulo, Brazil. The bees arrive in California in 1985 with permanent colonies in Texas by 1990, having spread through Mexico and Central America and Trinidad. The insects have moved into Utah, Georgia, and Tennessee and as far south as northern Argentina from Brazil.

1957 Giant ramshorn snail, a member of the apple snail family and native to central and northern South America, was first discovered in Coral Gables and became established in Florida. It turned up in the San Marcos River in Texas in 1981 and is also in the Comal River.

1958 Charles S. Elton publishes his pathbreaking *Ecology of Invasion by Plants and Animals.*

1960 Blue tilapia is widely disseminated in the United States from bait-bucket and fish-farm releases and soon establishes reproducing populations.

1962 David Wingate helps organize rat eradication from the Castle Harbour Islands Nature Reserve, including Nonsuch Island, in eastern Bermuda. The aim is to restore endemic plants and animals, including the long-believed-extinct Bermuda petrel. There are 34 invasive species on Bermuda.

1962 Red-rimmed melania, a freshwater snail native to tropical Africa and southern Asia, becomes established through the aquarium trade in many watersheds in Texas and exists in at least 11 additional states. It hosts a parasite that affects the gills of native fish.

1963 Chinese grass carp are imported into the United States as a possible, but controversial, biological control for hydrilla. During the following decade more than 30 producers raised fish in eight states. Diploid (fertile) carp have been stocked in about 40 states.

1970s Asian carp species (bighead, silver, and grass) are housed in the Mississippi watershed to clean commercial catfish ponds. Some escape and populate tributaries.

1972 Alfred W. Crosby writes *The Columbian Exchange* and in it coins the term "biological imperialism."

1980s Some 1,500 freshwater fish species are regularly imported into the United States for the aquarium trade, together with 100 or more species of aquatic plants.

1981 First legal importation and release of diploid (fertile) grass carp into Lake Conroe and Lewis Creek Reservoir, Texas.

About 1985 Red lionfish are likely released from aquaria off eastern Florida. They turn up on the Flower Garden Banks, 100 miles south of Galveston, Texas, in 2011.

1987 Infectious hematopoietic necrosis, native among juvenile salmon-type species along North America's Pacific Coast, is found in France and later in Belgium, Italy, Germany, Japan, and Taiwan.

1988 Zebra mussel, a shellfish native to the Black Sea basin, is found in Lake St. Clair, Canada, and arrived in waters of US Great Lakes two years later. It continues to spread in United States; was first documented in Lake Texoma, Texas, in 2009; and continues to infest lakes.

1989 Brown mussel, native to South America, is discovered on a jetty in Corpus Christi, Texas.

1989 Mediterranean fruit fly threatens the California citrus industry and results in massive spraying. There are five subsequent infestations (1995–2007) and others in Florida.

1990 Texas Parks and Wildlife requires a permit for the possession of carp species.

1990 Spiny daphnia, a water flea common in Africa, Asia, and Australia, is found in several places in Central and East Texas. It is not certain how the species got there.

1991 Nonindigenous Aquatic Species Program is established in the US Fish and Wildlife Service and moves to the United States Geological Survey Invasive Species Program in 1996.

1994 Invasive Species Specialist Group is established under sponsorship of the Species Survival Commission within the IUCN (now World Conservation Union). Based in Auckland, New Zealand, the group has 196 core members from 40 nations and a global network of 2,000 or more conservation practitioners. It established a Global Invasive Species database, including "100 of the World's Worst Invasive Alien Species," and since 1995 has published a biannual *Aliens* newsletter. There are three regional sections: North America, Europe, and South Asia.

1995 Half of the 150 species of alien plants and animals, which have known releases in and around San Francisco Bay, had been established since 1960.

1995 Giant salvinia, native to southeastern Brazil, floats in a small pond in Colleton County, South Carolina; within 20 years it appeared in 12 additional states (including Texas in 1998) as well as Puerto Rico and the US Virgin Islands.

1996 Asian long-horned beetle, a wood borer from China and Korea, is discovered in New York City. It kills thousands of hardwood trees in the Northeast, including Canada, and threatens 48 million acres of wood and forest, mostly in the Midwest.

By 1996 Rainbow trout, native to the western United States, have been planted successfully in more than 100 nations and islands. The US Bureau of Fisheries introduced the trout to Texas in 1882. It survives in West Texas and seasonally in colder water flows below dams on the Guadalupe and Brazos Rivers.

1997 Global Invasive Species Program is set up by a consortium of United Nations Environment Programme, International Union for Conservation of Nature, and Scientific Committee on Problems of the Environment personnel in response to the accelerated expansion of world trade.

1998 President William Clinton signs Executive Order 13112, coordinating and enhancing the federal government's abilities to minimize economic, public health, and ecological difficulties caused by invasive species.

Late 1990s to early 2000s Approximately 240 species of nonnative plant pathogens have become established in the United States; crop pests include Asian sorghum ergot (1998), soybean aphid (2001), and soybean rust (2006).

2000 Australian spotted jellyfish grows numerous enough to be recorded between Florida and Texas and is clogging estuaries from Mobile Bay to the Mississippi River delta.

2000 Burmese python is officially recognized as reproducing in Florida: sightings increase exponentially, and estimates place 300,000 in South Florida. In 2012 it is banned from being imported into the United States.

2000 At least 70 nonnative fish species are established in the United States, 23 of which are swimming in Florida. All but three were introduced for the aquarium industry.

2000 About 6,000 freshwater fish species make up the world's ornamental fish trade, and 59 of the most popular species are raised in the United States.

2000– 2006 Some 1.5 billion animals arrive in the United States in more than 500,000 pet shipments; many are wild-caught.

2001 Regional Euro-Asian Biological Invasions Centre is founded to make information about invasive species available online. It is the publisher of *Management of Biological Invasions*.

2002 Horus Institute for Environmental Conservation and Development is established to fight invasive species in Brazil.

2002 Emerald ash borer, native to China, is first identified in southeastern Michigan, where it has started to infest native ash trees, killing millions in 31 US states; four adult beetles discovered near Karnack, Texas, in 2016.

2002 Tawny crazy ant from South America discovered in Pasadena, Harris County, Texas. It then infests 20 counties, including Hidalgo County, close to the Mexican border.

2004 US Department of Agriculture/Animal and Plant Health Inspection Service/Wildlife Services, National Wildlife Research Center, US Department of the Interior Fish and Wildlife Service, and Wildlife Damage Management Working Group of the Wildlife Society sponsor a Second National Invasive Rodent Summit in National Wildlife Research Center, Fort Collins, Colorado (it includes rodent-control programs in New Zealand).

2005 Food and Agriculture Organization of the United Nations reviews the global impact of invasive tree species on forests and forestry.

2005 The National Invasive Species Information Center is formed in the National Agricultural Library to meet the information needs of users, including the National Invasive Species Council. The Information Center collects updated information for both experts and the interested public.

2005 First Texas Invasive Plant and Pest Conference takes place in the Lady Bird Johnson Wildflower Center, Austin, Texas, with 199 attendees.

2005 The Invaders of Texas Program is inaugurated for volunteers to detect the arrival and dispersal of invasive species in Texas. Within the first five years the program hosts 59 workshops and trains more than 1,300 citizen scientists.

2006 White-nose syndrome, a fungal disease infecting hibernating bats in North America, breaks out in Schoharie County, New York. It spreads to kill more than seven million bats of 11 species in 31 states and five Canadian provinces. The fungus appears in Texas in 2017. There is no known remedy. Ecological services provided by bats are valued at billions of dollars per year.

2007 US Fish and Wildlife Service declares the silver, largescale silver, and black carp invasive species under the Lacey Act and regards them as threats to the Great Lakes system.

2008 Texas Invasive Plant and Pest Council is established and adopts California's sister council's criteria for categorizing nonnative plants that threaten wild lands.

2013 At least 141 nonnative species are established in the US Great Lakes (including 26 fish, 29 insects, 24 algae, and 59 plants).

2015 US Department of Agriculture draws up invasive species profiles for 50 land plants, 16 aquatic plants, 28 insects, and 2 other invertebrates; 26 aquatic animals and 4 other animal vertebrates; and 7 animal and 17 plant pathogens in the United States and territories. The agency also publishes a watch list of 19 additional organisms likely to become established.

2015 Texas possesses at least 67 land and 12 aquatic invasive plant species, 11 insects, 10 mammals, 4 birds, 7 fishes, and 11 mollusks and crustaceans.

2016 Nineteenth International Conference on Aquatic Invasive Species is held in Winnipeg, Canada. It is recognized as the most comprehensive international forum and involves 400 or more participants from more than 30 countries.

2017 Biennial North American Invasive Species Forum meets in Savannah, Georgia, to share knowledge about management, regulations, dispersal, and ongoing issues related to invasive species.

NOTES

Preface

1. The Cornell Lab of Ornithology, All about Birds, "Great-tailed Grackle," accessed November 4, 2017, https://www.allaboutbirds.org/guide/Great-tailed_Grackle/id. Details on Christian Moeller's art installation, *CAW*, can be found in Anne Bruno, "Austin's New Central Library," *Tribeza*, November 1, 2017, http://tribeza.com/austins-new-central-library/.

2. Christopher Middleton, "Japanese Knotweed," July 5, 2014, http://www.newsweek.com/2014/07/11/japanese-knotweed-driving-men-murder-257257.html.

3. CBS Austin, "Parakeet Nest Fire," June 6, 2017, http://cbsaustin.com/news/local/video-parakeet-nest-fire-knocks-out-power-to-1000-austin-energy-customers; and Calily Bien, "Austin Energy, Travis Audubon Meet Half-Way," KXAN, June 8, 2017, http://kxan.com/2017/06/08/austin-energy-travis-audubon-meet-half-way-on-how-to-manage-monk-parakeets/.

Introduction

1. Vilà Montserrat and Philip E. Hulme, "Non-native Species, Ecosystem Services, and Human Well-Being," in *Impact of Biological Invasions on Ecosystem Services*, Invading Nature, Springer Series in Invasion Ecology, ed. Vilà Montserrat and Philip Hulme (New York: Springer, 2017), 12:1–14, https://link.springer.com/chapter/10.1007/978-3-319-45121-3_1.

2. Rita S. W. Yam, Ko-Pu Huang, Hwey-Lian Hsieh, Hsing-Juh Lin, and Shou-Chung Huang, "An Ecosystem-Service Approach to Evaluate the Role of Non-native Species in Urbanized Wetlands," *International Journal of Environmental Research and Public Health* 12, no. 4 (2015): 3939; and Adam Lampert, Alan Hastings, Edwin D. Grosholz, Sunny L. Jardine, and James N. Sanchirico, "Optimal Approaches for Balancing Invasive Species Eradication and Endangered Species Management," *Science* 344, no. 6187 (2014): 1028–31.

3. Banu Subramaniam, *Ghost Stories for Darwin* (Urbana: University of Illinois Press, 2014), 122.

4. Damian Carrington, "The Anthropocene Epoch: Scientists Declare Dawn of Human-Influenced Age," *The Guardian*, August 29, 2016, https://www.theguardian.com/environment/2016/aug/29/declare-anthropocene-epoch-experts-urge-geological-congress-human-impact-earth.

5. Simon L. Lewis and Mark A. Maslin, "Defining the Anthropocene," *Nature* 519 (March 12, 2015): 171–80.

6. Alfred W. Crosby, *The Columbian Exchange* (Westport, CT: Greenwood Publishing, 1972); and Knowles A. Ryerson, "History and Significance of the Foreign Plant

Introduction Work of the United States Department of Agriculture," *Agricultural History* 7, no. 3 (1933): 111–12.

7. Mark Van Kleunen, Wayne Dawson, Franz Essi, Jan Pergi, Marten Winter, Ewald Weber, Holger Kreft, et al., "Global Exchange and Accumulation of Non-native Plants," *Nature* 525 (September 3, 2015): 100–103.

8. Christopher Lever, *They Dined on Eland*: *The Story of Acclimatisation Societies* (London: Quiller, 1992).

9. Robin Doughty, *The Eucalyptus*: *Natural and Commercial History of the Gum Tree* (Baltimore: Johns Hopkins University Press, 2000).

10. Walter B. Barrows, *The English Sparrow* (Passer domesticus) *in North America*: *Especially in Its Relations to Agriculture*, USDA, Division of Economic Ornithology and Mammalogy, Bulletin 1 (Washington, DC: Government Printing Office, 1889).

11. Ryerson, "History and Significance," 121–23; and US Department of Agriculture, Bureau of Plant Industry, Office of Foreign Plant Introduction, *Plant Introductions*: *Sixteenth Annual List*: *1927–1928*, https://archive.org/details/plantintroducti01926unit.

12. Michael E. Dorcas, John D. Willson, Robert N. Reed, Ray W. Snow, Michael R. Rochford, Melissa A. Miller, Walter E. Meshaka Jr., et al., "Severe Mammal Declines Coincide with Proliferation of Invasive Burmese Pythons in Everglades National Park," *Proceedings of the National Academy of Sciences* 109 (2012): 2418–22.

13. For a complete account, see Sherrie S. McLeRoy and Roy E. Renfro Jr., *Grape Man of Texas*: *The Life of T. V. Munson* (Austin: Eakin Press, 2004).

14. James R. Casey, "The Mediterranean Fruit Fly in California: Taking Stock," *California Agriculture* 46, no. 1 (1992): 12–17. See also USDA, Animal and Plant Health Inspection Service, *Mediterranean Fruit Fly Action Plan*, 2003, http://www .aphis.usda.gov/import_export/plants/manuals/domestic/downloads/medfly_action_ plan.pdf.

15. We are basing these definitions heavily on works by two government agencies: the National Invasive Species Council, "Annex V: Glossary of Terms," in *Management Plan*: *2016–2018* (Washington, DC: National Invasive Species Council, 2016), 41–42; and US Department of the Interior, Bureau of Land Management, "About Weeds and Invasive Species," accessed October 28, 2017, https://www.blm.gov/programs/natural-resources/ weeds-and-invasives/about.

16. Exec. Order No. 13112, Fed. Reg. 64, no. 25 (February 3, 1999), 6183–86.

17. Charles C. Mann, *1493*: *Uncovering the New World Columbus Created* (New York: Vintage, 2012).

18. David L. Erickson, Bruce D. Smith, Andrew C. Clarke, Daniel H. Sandweiss, and Noreen Tuross, "An Asian Origin for a 10,000-Year-Old Domesticated Plant in the Americas," *PNAS* 102, no. 51 (2005): 18315–20. Native domesticated animals in the New World include the llama, guinea pig, turkey, Muscovy duck, stingless bee, and cochineal. See D. L. Johnson and B. K. Swartz Jr., "Evidence for Pre-Columbian Animal Domestication in the New World," paper presented at the Central States Anthropological Society meetings, March 11, 1989, https://soar.wichita.edu/handle/10057/1824; Birgitta Wallace, "L'Anse aux Meadows National Historic Site," in *Archeology in America*: *Encyclopedia*, 4 vols., ed. Francis P. McManamon, Linda S. Cordell, Kent G. Lightfoot, and George R. Milner (Westport, CT: Greenwood Publishing, 2009) 178–83; and Nicole Boivin, Rémy

Crassard, and Michael Petraglia, eds., *Human Dispersals and Species Movement* (Cambridge: Cambridge University Press, 2017), 211.

19. Norton G. Miller, "The Genera of Meliaceae in the Southeastern United States," *Journal of the Arnold Arboretum* 71 (1990): 470; and "André Michaux," accessed April 5, 2017, www.michaux.org/. For early Texas sightings, see Jean Louis Berlandier, *Journey to Mexico during the Years 1826 to 1834*, trans. Sheila M. Ohlendorf, Josette M. Bigelow, and Mary M. Standifer (Austin: Texas State Historical Association, 1980), 320; *A Visit to Texas: Being the Journal of a Traveller through Those Parts Most Interesting to American Settlers*, facsimile of the 1st ed. of 1834 (Austin: Steck, 1952), 39; and Viktor Bracht, *Texas in 1848*, trans. Charles Frank Schmidt (San Antonio: Naylor Printing, 1931), 38.

20. I. Jarić and G. Cvijanović, "The Tens Rule in Invasion Biology," *Environmental Management* 50, no. 6 (2012): 979–81.

21. Mark Williamson and Alastair Fitter, "The Varying Success of Invaders," *Ecology* 77, no. 6 (1996): 1662.

22. Van Kleunen et al. "Global Exchange," 100–103.

23. Union of Concerned Scientists, "Invasive Species Texas," 3, accessed October 6, 2016, http://www.texasinvasives.org/resources/publications/UOC_texasinvasives.pdf.

24. US Customs and Border Protection, news release, November 24, 2014, https://www.cbp.gov/newsroom/local-media-release/south-texas-us-customs-and-border-protection-agriculture-specialists; and https://mazatlantoday.net/mexican_federal_highway_40_40D.html, accessed November 6, 2017.

25. Ed Yong, "Japanese Animals Are Still Washing Up," *The Atlantic*, September 28, 2017, https://www.theatlantic.com/science/archive/2017/09/japanese-animals-are-still-washing-up-in-america-after-the-2011-tsunami/541347/.

26. Niels Aalund and William Fitzgibbon, "A Computational Modeler's Tour of the Port of Houston," in *Computational Methods and Models for Transport*, ed. P. Diez, P. Neittaanmäki, J. Periaux, T. Tuovinen, and O. Bräysy (New York: Springer International Publishing, 2018), 17–29, https://link.springer.com/chapter/10.1007%2F978-3-319-54490-8_2; and BlankRome Publications, "US Ballast Water Compliance Challenges Now That IMO's Ballast Water Convention Has Been Ratified," *Maritime*, no. 10 (September 2016), http://www.blankrome.com/index.cfm?contentID=37&itemID=4029.

Chapter 1

1. See "Source of the English Sparrow," accessed April 18, 2015, http://nyarc.org/arcade/brooklyn/articles/The_Source_of_the_English_Sparrow_%5Bexcerpt%5D.pdf. The archive presents conflicting evidence over the details of introduction. See also Michael P. Moulton, Wendell P. Cropper Jr., Michael L. Avery, and Linda E. Moultin, "The Earliest House Sparrow Introductions to North America," *Biological Invasions* 12 (2010): 2955–58; and Walter B. Barrows, *The English Sparrow* (Passer domesticus) *in North America: Especially in Its Relations to Agriculture*. USDA, Division of Economic Ornithology and Mammalogy, Bulletin 1 (Washington, DC: Government Printing Office, 1889), 17–18.

2. Robin Doughty, *The English Sparrow in the American Landscape*, Research Paper 19, School of Geography (Oxford: University of Oxford, 1978), 10, citing *Forest and Stream* 1877.

3. Kristina Alexander, *Injurious Species Listings under the Lacey Act: A Legal Briefing*, Congressional Research Service, August 1, 2013, 4, http://nationalaglawcenter.org/wp-content/uploads/assets/crs/R43170.pdf.

4. Ibid., appendix A, 5, 19–22.

5. W. Rhodes, extract from *Forest and Stream*, cited in Barrows, *The English Sparrow*, 305. See also Redpath Museum, "House Sparrow," Quebec Biodiversity Website, accessed April 16, 2018, http://redpath-museum.mcgill.ca/Qbp/birds/Specpages/housesparrow.htm.

6. Edward Brayton Clark, "The Unspeakable Sparrow," *Outing* 37 (1901): 408–411, quote on 405.

7. Pierre Acobas, "Shakespeare's Ornithology," accessed April 12, 2014, http://www.acobas.net/teaching/shakespeare/masters/.

8. Rhodes, cited in Barrows, *The English Sparrow*, 98.

9. Barrows, *The English Sparrow*, 80.

10. Ibid., 82, 84, 85, 86, 197–99, 358–88.

11. Ibid., 99. See also A. K. Fisher, "Walter Bradford Barrows," *The Auk* 42 (January 1925): 1–14.

12. Barrows, *The English Sparrow*, 315, citing *Forest and Stream*, August 6, 1885.

13. Barrows, *The English Sparrow*, 37.

14. Ibid., 319.

15. Ibid., 39.

16. Stanley D. Casto, "House Sparrow," in *Handbook of Texas Online*, accessed April 23, 2015, http://www.tshaonline.org/handbook/online/articles/tbh02, uploaded on June 15, 2010, published by the Texas State Historical Association.

17. David Sibley, "House Sparrow—'New' for North America?," *Sibley Guides*, February 19, 2008, http://www.sibleyguides.com/2008/02/house-sparrow-new-for-north-america/.

18. J. Denis Summers-Smith, *The House Sparrow* (London: Collins, 1963); and Ted R. Anderson, *Biology of the Ubiquitous House Sparrow: From Genes to Populations* (Oxford: Oxford University Press, 2006).

19. Ted Gup, "100 Years of Starlings," *New York Times*, September 1, 1990.

20. David Pimentel, Rodolfo Zuniga, and Doug Morrison, "Update on the Environmental and Economic Costs Associated with Alien-Invasive Species in the United States," *Ecological Economics* 52 (2005): 273–88; and Ron J. Johnson and James F. Glahn, "Starling Management in Agriculture," *Other Publications in Wildlife Management*, Paper 37 (1998), http://digitalcommons.unl.edu/cgi/viewcontent.cgi?article=1036&context=icwdmother.

21. "Lockheed L-188 Electra," Flight Safety Foundation database, Aviation Safety Network, October 4, 1960, https://aviation-safety.net/database/record.php?id=19601004-0.

22. Celebrate Boston, Category Disasters, "Worst Bird Strike, 1960," http://www.celebrateboston.com/disasters/logan-electra-bird-strike-1960.htm; Scott C. Barras, Sandra E. Wright, and Thomas E. Seamans, "Blackbird and Starling Strikes to Civil Aircraft in the United States, 1990–2001," USDA, *National Wildlife Research Center—Staff Publications*, Paper 200, 2003, http://digitalcommons.unl.edu/cgi/viewcontent.cgi?article=1194&context=icwdm_usdanwrc; and John R. Allan, "The Costs of Bird Strikes and Bird Strike

Prevention," *Human Conflicts with Wildlife: Economic Considerations*, Paper 18, 2000, http://www.birdstrike.it/birdstrike/file/images/file/The%20costs%20°f%20bird%20strikes.pdf.

23. George M. Linz, H. Jeffrey Homan, Shannon M. Gaulker, Linda B. Penry, and William J. Bleier, "European Starlings: A Review of an Invasive Species with Far-Reaching Impacts," Paper 24 in *Managing Vertebrate Invasive Species: Proceedings of an International Symposium*, ed. G. W. Witmer, W. C. Pitt, and K. A. Fagerstone (Fort Collins, CO: USDA/APHIS Wildlife Services, National Wildlife Research Center, 2007), 378–86; and Michael L. Avery and Eric A. Tillman, "Alien Birds in North America—Challenges for Wildlife Managers," Paper 87 in *Wildlife Damage Management Conferences—Proceedings of the 11th Wildlife Damage Management Conference*, ed. D. L. Nolte and K. A. Fagerstone (Lincoln: University of Nebraska, Lincoln, 2005), 82–89, https://digitalcommons.unl.edu/icwdm/.

24. Harry C. Oberholser, *Bird Life of Texas* (Austin: University of Texas Press, 1974), 2:698.

25. Robert C. Tweit, "The European Starling," in *The Texas Breeding Bird Atlas*, Texas A&M AgriLife Extension, 2005, http://txtbba.tamu.edu/species-accounts/european-starling/.

26. Charles Brown, "The Impact of Starlings on Purple Martin Populations in Unmanaged Colonies," *American Birds* 35, no. 3 (1981): 266–68.

27. Bob Buskas, "Martin Landlord," accessed April 23, 2015, http://www.wtc.ab.ca/northernskys/starling.htm.

28. "SREH Entrances," accessed April 12, 2018, http://www.chuckspurplemartinpage.com/srehentr.htm.

Chapter 2

1. John Klorer, "The Water Hyacinth Problem," *Journal of the Association of Engineering Societies* 42 (1909): 48.

2. Wm. T. Penfound and T. T. Earle, "The Biology of the Water Hyacinth," *Ecological Monographs* 18, no. 4 (1948): 449–50.

3. Ibid., 469.

4. Jon Mooallem, "American Hippopotamus," *The Atavist* 32 (December 2013), pt. 1, chap. 2, https://read.atavist.com/american-hippopotamus.

5. Ibid., pt. 1, chap. 2; pt. 3, chap. 12.

6. Paul P. Popenoe, "Meat Production in Swamps," *Journal of Heredity* 5 (1914): 34.

7. S. K. Eltringham, *The Hippos: Natural History and Conservation* (London: Academic Press, 1999), 3, 74, 78–82.

8. James N. Gowanloch, "The Economic Status of the Water-Hyacinth," *Louisiana Conservationist* 2 (1944): 3. The World's Fair introduction, though a commonly held theory and often cited in the literature, has its shortcomings. For critique and competing theories, see Donald C. Schmitz, Brian V. Nelson, Larry E. Nall, and Jeffrey D. Schardt, "Exotic Aquatic Plants in Florida: A Historical Perspective and Review of the Present Aquatic Plant Regulation Program," in *Proceedings of the Symposium on Exotic Pest Plants*, ed. T. D. Center, R. T. Doren, R. L. Hofstetter, R. L. Myers, and L. D. Whiteaker (Denver: US Department of the Interior / National Park Service, 1991), 306.

9. The River and Harbor Act of 1899, which included funds for water hyacinth removal in Florida and Louisiana, was amended on June 13, 1902, to include navigable

waters in the state of Texas. For hyacinth distribution in the first decade of the twentieth century in Texas, see Klorer, "Water Hyacinth Problem," 47.

10. Earl W. Chilton II, director of the Aquatic Habitat Enhancement Program, Texas Parks and Wildlife, pers. comm., 2015. See also his work with Flynt Houston and Howard Elder, *Nuisance Aquatic Vegetation Control in 2004*, Texas Parks and Wildlife Department, Management Data Series No. 239, 2006. See also Texas Water Conservation Association, "Texas Water Day 2007," accessed October 7, 2015, www.texaswca.com/texas-water-day/2007/4_Invasives.pdf; and Thad Sitton, *Caddo: Visions of a Southern Cypress Lake* (College Station: Texas A&M University Press, 2015), 156.

11. C. D. Sculthorpe, *The Biology of Aquatic Vascular Plants* (New York: St. Martin's Press, 1967), 461.

12. Yuan-Ye Zhang, Da-Yong Zhang, and Spencer C. H. Barrett, "Genetic Uniformity Characterizes the Invasive Spread of Water Hyacinth (*Eichhornia crassipes*), a Clonal Aquatic Plant," *Molecular Ecology* 19 (2010): 1775.

13. Brij Gopal and K. P. Sharma, *Water-Hyacinth* (Eichhornia crassipes), *the Most Troublesome Weed of the World* (Delhi: Hindasia, 1981).

14. Gowanloch, "Economic Status," 3.

15. Zhang, Zhang, and Barrett, "Genetic Uniformity," 1782.

16. Penfound and Earle, "Biology," 454; Gowanloch, "Economic Status," 3; and Klorer, "Water Hyacinth Problem," 48.

17. Penfound and Earle, "Biology," 453; and Richard Joel Russell, "Flotant," *Geographical Review* 32, no. 1 (1942): 97.

18. Spencer C. H. Barrett, "Waterweed Invasions," *Scientific American* 260 (1989): 90–91.

19. T. D. Center, "Waterhyacinth," in *Biological Control of Invasive Plants in the United States*, ed. Eric M. Coombs, Janet K. Clark, Gary L. Piper, and Alfred F. Cofrancesco Jr., (Corvallis: Oregon State University Press, 2004), 404.

20. P. E. Ndimele, C. A. Kumolu-Johnson, and M. A. Anetekhai, "The Invasive Aquatic Macrophyte, Water Hyacinth (*Eichhornia crassipes* [Mart.] Solm-Laubach: Pontedericeae): Problems and Prospects," *Research Journal of Environmental Sciences* 5 (2011): 513.

21. Larry D. Hodge, "Invasion of the Water Snatchers," *Texas Parks and Wildlife*, March 2015, https://www.tpwmagazine.com/archive/2015/mar/ed_2_invasives/.

22. C. E. Timmer and L. W. Weldon, "Evapotranspiration and Pollution of Water by Water Hyacinth," *Hyacinth Control Journal* 6 (1967): 34.

23. Klorer, "Water Hyacinth Problem," 34.

24. United States Army, Report of the Chief of Engineers, *Removing Water Hyacinth from Waters in Louisiana and Texas*, *U. S. War Department, Annual Reports 1909*, 9 vols. (Washington, DC: Government Printing Office, 1909), 5:466–68; and Klorer, "Water Hyacinth Problem," 34–35.

25. Maricela Martínez Jiménez, "Progress on Water Hyacinth (*Eichhornia crassipes*) Management," Food and Agriculture Organization Plant Production and Protection Paper 120, Addendum 1, 2003, http://www.fao.org/docrep/006/y5031e/y5031e0c.htm.

26. Gowanloch, "Economic Status," 6. See also William E. Wunderlich, "Mechanical Hyacinth Destruction," *Military Engineer* 30, no. 169 (1938): 5–8.

27. US Geological Survey, "Nutria, Eating Louisiana's Coast," USGS Fact Sheet 020-00, June 2000, updated April 20, 2001; and Shane K. Bernard, "M'sieu Ned's Rat?

Reconstructing the Origin of Nutria in Louisiana: The E. A. McIlhenny Collection, Avery Island, Louisiana," *Louisiana History*: *The Journal of the Louisiana Historical Association* 43, no. 3 (2002): 287, 290–91.

28. America's Wetland Foundation, "Nutria," accessed October 10, 2017, http://www.americaswetlandresources.com/wildlife_ecology/plants_animals_ecology/animals/mammals/nutria.html; and Sitton, *Caddo*, 109, 156–57.

29. Earl L. Atwood, "Life History Studies of Nutria, or Coypu, in Coastal Louisiana," *Journal of Wildlife Management* 14, no. 3 (1950): 251; and Wendell G. Swank and George A. Petrides, "Establishment and Food Habits of the Nutria in Texas," *Ecology* 35, no. 2 (1954): 173.

30. M. D. Haigh, "The Use of Manatees for the Control of Aquatic Weeds in Guyana," *Irrigation and Drainage Systems* 5, no. 4 (1991): 342; and Harvey Rice, "Rare Manatee Sighting in Galveston," *Houston Chronicle*, October 15, 2012, http://www.chron.com/news/houston-texas/article/Rare-manatee-sighting-in-Galveston-3924028.php.

31. Center, "Waterhyacinth," 407; Philip W. Tipping, Alejandro Sosa, Eileen N. Pokorny, Jeremiah Foley, Don C. Schmitz, Jon S. Lane, Leroy Rodgers, et al., "Release and Establishment of *Megamelus scutellaris* (Hemiptera: Delphacidae) on Waterhyacinth in Florida," *Florida Entomologist* 97, no. 2 (2014): 804–6; and Eric M. Coombs, Hans Radtke, and Thomas Nordblom, "Economic Benefits of Biological Control," in Coombs et al., *Biological Control of Invasive Plants in the United States*, 124.

32. B. C. Wolverton and Rebecca C. McDonald, "Water Hyacinth (*Eichhornia crassipes*) Productivity and Harvesting Studies," *Economic Botany* 33, no. 1 (1979): 1–10; and US Environmental Protection Agency, *Design Manual*: *Constructed Wetlands and Aquatic Plant Systems for Municipal Wastewater Treatment* (Cincinnati, OH: Center for Environmental Research Information, 1988), 61–77.

33. Garry Hamilton, "Striking a Deal with the Weed from Hell," *Conservation Magazine* 15, no. 1 (2014), http://conservationmagazine.org/2014/03/water-hyacinth-in-kings-bay/.

34. Ibid.

35. Ibid.

36. Ibid.; and Robert Knight, pers. comm., 2016.

37. Washington State Department of Ecology, "General Information about Hydrilla," accessed March 30, 2015, http://www.ecy.wa.gov/programs/wq/plants/weeds/hydrilla.html.

38. Kenneth A. Langeland, "*Hydrilla verticillata* (L.F.) Royle (Hydrocharitaceae), 'the Perfect Aquatic Weed,'" *Castanea* 61, no. 3 (1996): 297.

39. Leeann M. Glomski and Michael D. Netherland, "Does Hydrilla Grow an Inch per Day? Measuring Short-Term Changes in Shoot Length to Describe Invasive Potential," *Journal of Aquatic Management* 50 (2012): 56.

40. Global Invasive Species Database, "*Hydrilla verticillata*," accessed April 28, 2018, http://issg.org/database/species/ecology.asp?si=272&fr=1&sts=sss&lang=EN; and D. C. Schmitz, J. D. Schardt, A. J. Leslie, F. A. Dray Jr., J. A. Osborne, and B. V. Nelson, "The Ecological Impact and Management History of Three Invasive Alien Aquatic Plant Species in Florida," in *Biological Pollution*: *The Control and Impact of Invasive Exotic Species*, ed. Bill N. McKnight (Indianapolis: Indiana Academy of Science, 1993), 179.

41. Larry D. Hodge, "Exotic Aquatics . . . Manna from Heaven or Weeds from Hell?," *Texas Fish & Game*, April 2004, https://tpwd.texas.gov/fishboat/fish/didyouknow/inland/exotic_aquatics.phtml.

42. Seth Robbins, "Drownings Spike along Rio Grande," *Austin American-Statesman*, March 30, 2015; and Gene Ingalsbe, "Aquatic Weed Researcher Dr. Lyle Weldon Drowns," *Weeds Trees and Turf* 9, no. 3 (1970): 49, http://archive.lib.msu.edu/tic/wetrt/article/1970mar.pdf. See also Schmitz et al., "Exotic Aquatic Plants in Florida," 311.

43. Quoted in Hodge, "Exotic Aquatics"; and Langeland, "*Hydrilla verticillata*," 298–99.

44. Christopher D. K. Cook and Ruth Lüönd, "A Revision of the Genus *Hydrilla* (Hydrocharitaceae)," *Aquatic Botany* 13 (1982): 490; J. K. Balciunas, M. J. Grodowitz, A. F. Cofrancesco, and J. F. Shearer, "Hydrilla," in *Biological Control of Invasive Plants in the Eastern United States*, ed. Roy Van Driesche, Suzanne Lyon, Bernd Blossey, Mark Hoddle, and Richard Reardon, Publication FHTET-2002-04 (Morgantown, WV: US Department of Agriculture, Forest Service, Forest Health Technology Enterprise Team, 2002), 92–93; and Schmitz et al., "Exotic Aquatic Plants in Florida," 311.

45. Paul T. Madeira, Thai K. Van, Kerry K. Steward, and Raymond J. Schnell, "Random Amplified Polymorphic DNA Analysis of the Phenetic Relationships among World-Wide Accession of *Hydrilla verticillata*," *Aquatic Botany* 59 (1997): 231; and Washington State Department of Ecology, "General Information."

46. Earl Chilton, director of Aquatic Habitat Enhancement Program, Texas Parks and Wildlife Department, pers. comm., 2015.

47. Tobias A. Schmid, J. P. Cuda, and G. E. Macdonald, "Performance of Two Established Biological Control Agents on *Hydrilla* Genotypes Susceptible and Resistant to Fluridone Herbicide," *Journal of Aquatic Plant Management* 48 (2010): 102.

48. R. D. Doyle and R. M. Smart, "Effects of Drawdowns and Dessication [*sic*] on Tubers of Hydrilla, an Exotic Aquatic Weed," *Weed Science* 49, no. 1 (2001): 137.

49. Balciunas et al., "Hydrilla," 101.

50. John Cassani, Scott Hardin, Vince Mudrak, and Paul Zajicek, *A Risk Analysis Pertaining to the Use of Triploid Grass Carp for the Biological Control of Aquatic Plants* (Tallahassee: Florida Department of Environmental Protection and Florida Department of Agriculture and Consumer Services, 2008), 17.

51. Shannon Tompkins, "Grass Carp Keep Hydrilla in Check, but Can Kill Vegetation," *Houston Chronicle*, July 14, 2011. See also Texas Parks and Wildlife news release, "Anglers to Target Grass Carp on Lake Conroe," June 20, 2011. https://tpwd.texas.gov/newsmedia/releases/?req=20110620a.

52. Earl W. Chilton II, Mark A. Webb, and Richard A. Ott Jr., "Hydrilla Management in Lake Conroe, Texas: A Case History," in *Balancing Fisheries Management and Water Uses for Impounded River Systems*, ed. Micheal S. Allen, Steve Sammons, and Michael J. Maceina, American Fisheries Society Symposium 62 (2008): 254.

53. Earl W. Chilton II and Stephan J. Magnelia, "Use of an Incremental Triploid Grass Carp Stocking Strategy for Maintaining Vegetation Coverage in a Riverine Texas Reservoir," in Allen, Sammons, and Maceina, *Balancing Fisheries Management and Water Uses for Impounded River Systems*, 545.

54. Ibid., 552.

55. Mike Bleier, "President's Update—December 19, 2014," Lake Conroe Association, accessed October 23, 2015, www.lakeconroeassociation.com/presidents-update-december-19-2014/.

56. Ricardo Gandara, "Survey Finds No Hydrilla in Lake Austin," *Austin American-Statesman*, October 23, 2013.

57. Austin, Texas, City of Austin Watershed Protection Department, "Sterile Carp Release in Lake Austin for Hydrilla Control: Release Date: May 2, 2013," accessed April 17, 2015, http://www.austintexas.gov/article/sterile-carp-released-lake-austin-hydrilla-control.

58. Mike Leggett, "Weather Making It Tough to Find, Catch White Bass," *Austin American-Statesman*, February 16, 2015, https://www.mystatesman.com/sports/leggett-weather-making-tough-find-catch-white-bass/IYLE9KAIPOAz1dX76kzrOO/.

59. Gandara, "Survey Finds No Hydrilla."

60. Nick Visser, "Eat the Enemy: The Delicious Solution to Menacing Asian Carp," *Huffington Post*, December 15, 2014, http://www.huffingtonpost.com/2014/12/15/eat-the-enemy-asian-carp_n_6324896.html; and Eat the Invaders, "Asian Carp," February 15, 2012, http://eattheinvaders.org/asian-carp/.

Chapter 3

1. Asher Elbein, "War on Feral Hogs in Texas," *Texas Observer*, July 24, 2017.

2. Reeve Hamilton, "Texplainer: How Real Is the Feral Hog Menace?," *Texas Tribune*, April 4, 2011; B. C. West, A. L. Cooper, and J. B. Armstrong, "Managing Wild Pigs: A Technical Guide," *Human-Wildlife Interactions Monograph* 1 (2009): 1–55; and Billy Higginbotham, "Feral Hogs: Questions and Answers," October 2017, hoasites.goodwintx.com/Portals/157/Feral%20Hogs%20information.pdf. Higginbotham suggests that El Paso County will be the last holdout against this wild pig swarm.

3. Erica Goode, "When One Man's Game Is Also a Marauding Pest," *New York Times*, April 28, 2013; John Morthland, "A Plague of Pigs in Texas," *Smithsonian Magazine*, January 2011, http://www.smithsonianmag.com/science-nature/A-Plague-of-Pigs-in-Texas.html#ixzz1NYronPRi; and United States Department of Agriculture, APHIS, "USDA Announces a $20 Million Effort to Reduce Damage Caused by Feral Swine," news release, April 2, 2014, https://www.aphis.usda.gov/newsroom/2014/04/pdf/feral_swine.pdf.

4. Alfred W. Crosby, *The Columbian Exchange*: *Biological and Cultural Consequences of 1492* (Westport, CT: Greenwood: 1972), 77.

5. John J. Mayer and I. Lehr Brisbin, *Wild Pigs in the United States: Their History, Comparative Morphology, and Current Status* (Athens: University of Georgia Press, 2008), 21, passim; "Feral-Hog Questions and Answers," accessed March 23, 2015, https://www.youtube.com/watch?v=OtUOaDro1vs; and Texas A&M University, AgriLife and Extension, "Feral Hog Population: Growth, Density and Harvest in Texas," 2012, http://www.invasivespecies.wa.gov/documents/squealonpigs/FeralHogPopGrowthDensity&HarvestinTX.pdf.

6. Rick Taylor, *The Feral Hog in Texas*, Texas Parks and Wildlife, accessed April 12, 2013, http://tpwd.texas.gov/huntwild/wild/nuisance/feral_hogs/.

7. Brian McCombie, "Swine Story," *Game & Fish*, April 23, 2013, http://www.gameandfishmag.com/hunting/history-of-feral-hogs/.

8. Kathleen Phillips, "Busting Feral Hog Myths," *AgriLife Today*, March 24, 2011, http://today.agrilife.org/2011/03/24/busting-feral-hog-myths/.

9. Jared B. Timmons, Billy Higginbotham, Roel Lopez, James C. Cathey, Janell Mellish, Jonathan Griffin, Aaron Sumrall, et al., *Feral Hog Population Growth, Density, and Harvest in Texas*, Texas A&M, AgriLife, August 2012, 2, http://www.invasivespecies.wa.gov/documents/squealonpigs/FeralHogPopGrowthDensity&HarvestinTX.pdf.

10. Mike Leggett, "Helicopters? That's Going the Whole Hog," *Austin American-Statesman*, May 22, 2011. Hogs for a Cause is a self-professed sportsman's ministry that seeks to eliminate groups of feral hogs in Texas by providing hog meat to local people, notably wounded military personnel. See "Hog Services," Hogs for a Cause, accessed April 10, 2018, http://www.hogsforacause.org/hog-eradication. Darby Kendall noted that in 2014, heli-hunters donated 38,000 pounds of hog meat to Bell County residents. See Darby Kendall, "Nonprofit Group Unites around Food, Faith and Feral Hogs," *Reporting Texas,* April 14, 2015, reportingtexas.com/nonprofit-group-unites-around-food-faith-and-feral-hogs/; see also Chris Waddington, "Hogs for the Cause 2015 Drew Record 30,000 Fans to New Orleans Barbecue Fest," *Times-Picayune*, March 30, 2015, http://www.nola.com/festivals/index.ssf/2015/03/hogs_for_the_cause_2015_raised.html.

11. Clara O'Rourke, "New Plans Emerge to Control Hog Population," *Austin American-Statesman*, January 22, 2013.

12. Jeffrey T. Wilcox, Erik T. Aschehoug, Cheryl A. Scott, and Dirk H. Van Vuren, "A Test of the Judas Technique as a Method for Eradicating Feral Pigs," *Transactions of the Western Section of the Wildlife Society* 40 (2004): 120–26.

13. "Hunting with Hog Dogs," accessed April 10, 2018, http://www.blackmouthcur.com/HA03.htm; see also "Welcome to www.BlackMouthCur.com," accessed April 10, 2018, http://www.blackmouthcur.com/index.htm#TopOfPage; and John Morthland, "A Plague of Pigs in Texas," *Smithsonian Magazine*, January 2011, https://www.smithsonianmag.com/science-nature/a-plague-of-pigs-in-texas-73769069/.

14. Morgan Smith, "Feral Hog Poison in Texas on Hold," *Texas Tribune*, April 25, 2017.

15. Australia, Department of the Environment, Water, Heritage, and the Art, Veterinary Services Division, Institute of Medical and Veterinary Science, *Assessing the Humaneness and Efficacy of a New Feral Pig Bait in Domestic Pigs*, Study PC0409, March 2010, 10, https://www.environment.gov.au/system/files/resources/091b0583-f35c-40b3-a530-f2e0c307a20c/files/pigs-imvs-report.pdf.

16. Ralph Winingham, "Nighttime Hunters Help Control Feral Hog Population," *MYSA*, San Antonio's Home Page, May 20, 2011, http://www.mysanantonio.com/sports/outdoors/article/Nighttime-hunters-help-control-feral-hog-1387491.php#ixzz1NIICj4Ie.

17. Brooke Crum, "Special Report: Is Poison the Answer to Feral Hog Problem?," *Abilene Reporter-News*, April 3, 2017, http://www.reporternews.com/story/news/local/2017/03/04/poison-answer-feral-hog-problem/98705028/; and Stephen Lapidge, Jason Wishart, Michelle Smith, and Linton Staples, "Is America Ready for a Humane Feral Pig 'Toxin'?," Wildlife Damage Management Conference 13 (2009): 49–59.

18. Tyler Campbell, "Poison Bait May Help Control Wild Hogs," *Farm Show* 36, no. 4 (2012): 17, http://www.farmshow.com/a_article.php?aid=25789.

Chapter 4

1. Thomas Jefferson, "Summary of Public Service," The Thomas Jefferson Papers, Series 1, Library of Congress, September 1800, https://www.loc.gov/resource/mtj1.022_0449_0450/.

2. Troy C. Barrilleaux and James B. Grace, "Growth and Invasive Potential of *Sapium sebiferum* (Euphorbiaceae) within the Coastal Prairie Region: The Effects of Soil and Moisture Regime," *American Journal of Botany* 87 (2000): 1099.

3. David Hooper, "Chinese or Vegetable Tallow: Its Preparation, Uses and Composition," *Agricultural Ledger* (India) 2 (1904): 11.

4. Ellis's most notable discovery was the animal nature of coral.

5. Malcolm Bell III, *Some Notes and Reflections upon a Letter from Benjamin Franklin to Noble Wimberly Jones, October 7, 1772* (Darien, GA: Ashantilly Press, 1966), 3–6, 8; Franklin's letter to Noble Jones follows preface.

6. Stephen Elliott, *Sketch of the Botany of South Carolina and Georgia* (Charleston, SC: J. R. Schenck, 1824), 2:651.

7. Bell, *Some Notes*, 8; and Johann David Schöpf, *Reise durch einige der mittlern und südlichen vereinigten nordamerikanischen Staaten nach Ost-Florida und den Bahama-Inseln unternommen in den Jahren 1783 und 1784* (Erlangen, Germany: J. J. Palm, 1788), 271, 286.

8. F. N. Howes, "The Chinese Tallow Tree (*Sapium sebiferum* Roxb.)—a Source of Drying Oil," *Kew Bulletin* 4, no. 4 (1949): 575; G. S. Jamieson and R. S. McKinney, "Stillingia Oil," *Oil and Soap* 15, no. 11 (1938): 295; and Herbert W. Scheld, J. R. Cowles, C. R. Engler, R. Kleiman, and E. B. Schultz Jr., "Seeds of the Chinese Tallow Tree as a Source of Chemicals and Fuels," in *Fuels and Chemicals from Oilseeds: Technology and Policy Options*, ed. Eugene B. Schultz Jr. and Robert P. Morgan, American Association for the Advancement of Science Selected Symposium 91 (Boulder, CO: Westview Press, 1984), 85.

9. W. M. Potts and Don S. Bolley, "Analysis of the Fruit of the Chinese Tallow Tree in Texas," *Oil and Soap* 23 (1946): 316; and Flori Meeks, "Teas Left Green Legacy in Houston," *Houston Chronicle*, July 17, 2012, http://www.chron.com/news/article/Teas-left-green-legacy-in-Houston-3713653.php.

10. W. M. Potts, "The Chinese Tallow Tree as a Chemurgic Crop," *Chemurgic Digest* 5, no. 22 (1946): 373.

11. Notes on the collection label of a herbarium specimen (Robert Runyon 845, May 1, 1925) housed at The University of Texas at Austin's Billie L. Turner Plant Resources Center; L. H. Russell, W. L. Schwartz, and J. W. Dollahite, "Toxicity of Chinese Tallow Tree (*Sapium sebiferum*) for Ruminants," *American Journal of Veterinary Research* 30, no. 7 (1969): 1233; and Howes, "Chinese Tallow," 575.

12. Robert S. Gray, "Chinese Tallow Tree—a Four-Way Crop?," *Farm Journal* 74 (August 1950): 124.

13. Saara J. DeWalt, Evan Siemann, and William E. Rogers, "Geographic Distribution of Genetic Variation among Native and Introduced Populations of Chinese Tallow Tree, *Triadica sebifera* (Euphorbiaceae)," *American Journal of Botany* 98, no. 7 (2011): 1128–38.

14. Robert A. Vines, *Trees, Shrubs, and Woody Vines of the Southwest* (Austin: University of Texas Press, 1960), 622.

15. E. W. Eckey, *Vegetable Fats and Oils*, American Chemical Society Monograph Series (New York: Reinhold Publishing, 1954), 597.

16. Katherine A. Bruce, Guy N. Cameron, and Paul A. Harcombe, "Initiation of a New Woodland Type on the Texas Coastal Prairie by the Chinese Tallow Tree (*Sapium sebiferum* [L.] Roxb.)," *Bulletin of the Torrey Botanical Club* 122, no. 3 (1995): 216.

17. Bruce et al., "Initiation of a New Woodland Type," 223; Guy N. Cameron and Stephen R. Spencer, "Entomofauna of the Introduced Chinese Tallow Tree," *Southwestern Naturalist* 55, no. 2 (2010): 180; and Hsiao-Hsuan Wang, William E. Grant, Todd M. Swannack, Jianbang Gan, William E. Rogers, Tomasz E. Koralewski, James H. Miller, et al.,

"Predicted Range Expansion of Chinese Tallow Tree (*Triadica sebifera*) in Forestlands of the Southern United States," *Diversity and Distributions* 17, no. 3 (2011): 552–53, 559.

18. Jianbang Gan, James H. Miller, Hsiaohsuan Wang, and John W. Taylor Jr., "Invasion of Tallow Tree into Southern US Forests: Influencing Factors and Implications for Mitigation," *Canadian Journal of Forest Research* 39, no. 7 (2009): 1346.

19. Stephanie Flack and Elaine Furlow, "America's Least Wanted: A Lineup of the Country's Twelve Meanest Environmental Scoundrels," *Nature Conservancy Magazine* 46, no. 6 (1996): 23.

20. "Chinese Tallowtree, Early Detection & Distribution Mapping System," University of Georgia, Center for Invasive Species and Ecosystem Health, accessed August 16, 2017, http://www.eddmaps.org/.

21. Katherine A. Bruce, Guy N. Cameron, Paul A. Harcombe, and Greg Jubinsky, "Introduction, Impact on Native Habitats, and Management of a Woody Invader, the Chinese Tallow Tree, *Sapium sebiferum* (L.) Roxb.," *Natural Areas Journal* 17, no. 3 (1997): 256.

22. Ibid.

23. Jianwen Zou, William E. Rogers, and Evan Siemann, "Increased Competitive Ability and Herbivory Tolerance in the Invasive Plant *Sapium sebiferum*," *Biological Invasions* 10, no. 3 (2008): 291–302; and Jianwen Zou, William E. Rogers, Saara J. DeWalt, and Evan Siemann, "The Effect of Chinese Tallow Tree (*Sapium sebiferum*) Ecotype on Soil-Plant System Carbon and Nitrogen Processes," *Oecologia* 150, no. 2 (2006): 280.

24. Ian J. Renne, Timothy P. Spira, and W. C. Bridges Jr., "Effects of Habitat, Burial, Age and Passage through Birds on Germination and Establishment of Chinese Tallow Tree in Coastal South Carolina," *Journal of the Torrey Botanical Society* 128, no. 2 (2001): 109–19. *Wu* (in *wu-yau-shoe*) is Mandarin for "crow"; see Clarke Abel, *Narrative of a Journey in the Interior of China, and of a Voyage to and from That Country, in the Years 1816 and 1817; Containing an Account of the Most Interesting Transactions of Lord Amherst's Embassy to the Court of Pekin, and Observations on the Countries Which It Visited* (London: Longman, Hurst, Rees, Orme, and Brown, 1818), 177. Also see Warren C. Conway, Loren M. Smith, and James F. Bergan, "Avian Use of Chinese Tallow Seeds in Coastal Texas," *Southwestern Naturalist* 47, no. 4 (2002): 553; William Colson and Alan Fedynich, "Observations of Unusual Feeding Behavior of White-Winged Dove on Chinese Tallow," *Southwestern Naturalist* 61, no. 2 (2016): 133–35; and Ian J. Renne, Wylie C. Barrow Jr., Lori A. Johnson Randall, and William C. Bridges Jr., "Generalized Avian Dispersal Syndrome Contributes to Chinese Tallow Tree (*Sapium sebiferum*, Euphorbiaceae) Invasiveness," *Diversity and Distributions* 8, no. 5 (2002): 290.

25. Michael J. Baldwin, "Metabolizable Energy in Chinese Tallow Fruit for Yellow-Rumped Warblers, Northern Cardinals, and American Robins," *Wilson Journal of Ornithology* 120, no. 3 (2008): 528.

26. Wylie C. Barrow Jr. and Ian Renne, "Interactions between Migrant Landbirds and an Invasive Exotic Plant: The Chinese Tallow Tree," Texas Parks and Wildlife Department, *Flyway Newsletter* 8 (2001): 11.

27. Ian J. Renne, Sidney A. Gauthreaux Jr., and Charles A. Gresham, "Seed Dispersal of the Chinese Tallow Tree (*Sapium sebiferum* [L.] Roxb.) by Birds in Coastal South Carolina," *American Midland Naturalist* 144, no. 1 (2000): 211.

28. Evan Siemann, Wylie C. Barrow Jr., Lori A. Johnson Randall, and William C. Bridges, Jr., "Experimental Test for the Impacts of Feral Hogs on Forest Dynamics and Processes in the Southeastern US," *Forest Ecology and Management* 258, no. 5 (2009): 552.

29. Robert R. Pattison and Richard N. Mack, "Potential Distribution of the Invasive Tree *Triadica sebifera* (Euphorbiaceae) in the United States: Evaluating CLIMEX Predictions with Field Trials," *Global Change Biology* 14, no. 4 (2008): 820.

30. Jerome J. Howard, "Hurricane Katrina Impact on a Leveed Bottomland Hardwood Forest in Louisiana," *American Midland Naturalist* 168, no. 1 (2012): 61–62, 66; and Qiang Wang, Xingzhong Yuan, Hong Liu, Yuewei Zhang, Zhongli Cheng, and Bo Li, "Effect of Long-Term Winter Flooding on the Vascular Flora in the Drawdown Area of the Three Georges [*sic*] Reservoir, China," *Polish Journal of Ecology* 60, no. 1 (2012): 95–96, 102–4.

31. Hsiao-Hsuan Wang, William E. Grant, Jianbang Gan, William E. Rogers, Todd M. Swannack, Tomasz E. Koralewski, James H. Miller, et al., "Integrating Spread Dynamics and Economics of Timber Production to Manage Chinese Tallow Invasion in Southern U.S. Forestlands," *PLOS One* 7, no. 3 (2012): 8.

32. James B. Grace, "Can Prescribed Fire Save the Endangered Coastal Prairie Ecosystem from Chinese Tallow Invasion?," *Endangered Species Update* 15, no. 5 (1998): 74.

33. Bruce et al., "Introduction, Impact," 259.

34. Gregory S. Wheeler, M. Sedonia Steininger, and Susan Wright, "Quarantine Host Range of *Bikasha collaris*, a Potential Biological Control Agent of Chinese Tallowtree (*Triadica sebifera*) in North America," *Entomologia Experimentalis et Applicata* 163, no. 2 (2017): 184–96.

35. Frank C. Pellet, *American Honey Plants*, 5th ed. (Hamilton, IL: Dadant and Sons, 1976), 400.

36. Bernie Hayes, "The Chinese Tallow Tree—Artificial Bee Pasturage Success Story," *American Bee Journal* 119, no. 12 (1979): 848.

37. Meredith Hoag Lieux, "Dominant Pollen Types Recovered from Commercial Louisiana Honeys," *Economic Botany* 29, no. 1 (1975): 87, 95; and Scheld et al., "Seeds of the Chinese Tallow," 95.

38. Eugene B. Schultz Jr., William P. Darby, Harold M. Draper III, and Robert P. Morgan, "Novel Marginal-Land Oilseeds: Potential Benefits and Risks," in Schultz and Morgan, *Fuels and Chemicals from Oilseeds*, 23; and Susanne Retka Schill, "Franklin's Gift," *Biodiesel Magazine*, January 15, 2009, http://www.biodieselmagazine.com/articles/3158/franklin's-gift.

39. Scheld et al., "Seeds of the Chinese Tallow," 94.

40. L. H. Princen, "New Oilseed Crops on the Horizon," *Economic Botany* 37, no. 4 (1983): 489.

41. Yun Liu, Hong-Ling Xin, and Yun-Jun Yan, "Physicochemical Properties of Stillingia Oil: Feasibility for Biodiesel Production by Enzyme Transesterification," *Industrial Crops and Products* 30, no. 3 (2009): 436.

42. Scheld et al., "Seeds of the Chinese Tallow," 83.

43. Schill, "Franklin's Gift."

44. Schultz et al., "Novel Marginal-Land Oilseeds," 14.

45. Schill, "Franklin's Gift."

Chapter 5

1. Aldo Leopold, *A Sand County Almanac, and Sketches Here and There* (New York: Oxford University Press, 1949), 224–25.

2. Kevin Hultine and Tom Dudley, "*Tamarix* from Organism to Landscape," in *Tamarix: A Case Study of Ecological Change in the American West*, ed. Anna Sher and Martin F. Quigley (Oxford: Oxford University Press, 2013), 163.

3. T. W. Robinson, *Introduction, Spread, and Areal Extent of Saltcedar* (Tamarix) *in the Western States*, Geological Survey Professional Paper 491-A (Washington, DC: US Government Printing Office, 1965), A5.

4. Ibid.

5. Martin F. Quigley, "Botany and Horticulture of Tamarisk," in Sher and Quigley, *Tamarix*, 323.

6. John F. Gaskin and David J. Kazmer, "Introgression between Invasive Saltcedars (*Tamarix chinensis* and *T. ramosissima*) in the USA," *Biological Invasions* 11, no. 5 (2009): 1125, 1127.

7. W. D. Peck, *A Catalogue of American and Foreign Plants, Cultivated in the Botanic Garden, Cambridge, Massachusetts* (Cambridge, MA: Hilliard and Metcalf, 1818), 13; and Matthew K. Chew, "*Tamarix* Introduction, Naturalization, and Control in the United States, 1818–1952," in Sher and Quigley, *Tamarix*, 270.

8. J. F. Joor, "The Tamarisk Naturalized." *Bulletin of the Torrey Botanical Club* 6, no. 32 (1877): 166.

9. S. M. Mansfield, "Appendix S: Improvement of Rivers and Harbors in the State of Texas," in *Annual Report of the Chief of Engineers, US Army, to the Secretary of War, for the Year 1886*, part 2 (Washington, DC: US Government Printing Office, 1886), 1332.

10. Mark Alfred Carleton, "Adaptation of the Tamarisk for Dry Lands," *Science* 39, no. 1010 (1914): 693.

11. Donald J. Pisani, *Water and American Government: The Reclamation Bureau, National Water Policy, and the West, 1902–1935* (Berkeley: University of California Press, 2002), 2.

12. US Department of the Interior, Bureau of Reclamation, "Reclamation: Managing Water in the West," updated July 28, 2015, http://www.usbr.gov/main/about/.

13. Anna Sher, "Introduction to the Paradox Plant," in Sher and Quigley, *Tamarix*, 1.

14. William L. Graf, "Fluvial Adjustments to the Spread of Tamarisk in the Colorado Plateau Region," *Geological Society of America Bulletin* 89 (1978): 1494.

15. Lloyd H. Shinners, "Geographical Limits of Some Alien Weeds in Texas," *Texas Geographic Magazine* 12 (1948): 18; Robinson, "Introduction, Spread and Areal Extent," A1, A11; and Carleton, "Adaptation of the Tamarisk," 693.

16. W. H. Blackburn, R. W. Knight, and J. L. Schuster, "Saltcedar Influence on Sedimentation in the Brazos River," *Journal of Soils and Water Conservation* 37, no. 5 (1982): 298; and Daniel A. Auerbach, David M. Merritt, and Patrick B. Shafroth, "Tamarix, Hydrology, and Fluvial Geomorphology," in Sher and Quigley, *Tamarix*, 111.

17. Gail M. Drus, "Fire Ecology of *Tamarix*," in Sher and Quigley, *Tamarix*, 241.

18. Michelle K. Ohrtman and Ken D. Lair, "Tamarix and Salinity: An Overview," in Sher and Quigley, *Tamarix*, 138.

19. An acre-foot is the amount of water needed to cover one acre of surface to a depth

of one foot, or 325,851 gallons. Historically in US water-management circles, an acre-foot was considered the amount a suburban family household would use in a year. Recent conservation trends in the Southwest would allot the same amount to four or five families per year.

20. Chew, "*Tamarix* Introduction, Naturalization," 277; and Erika Zavaleta, "The Economic Value of Controlling an Invasive Shrub," *Ambio: A Journal of the Human Environment* 29, no. 8 (2000): 465, table 4.

21. Pamela L. Nagler and Edward P. Glenn, "Tamarisk: Ecohydrology of a Successful Plant," in Sher and Quigley, *Tamarix*, 78; and Tim Carlson, "The Politics of a Tree," in Sher and Quigley, *Tamarix*, 291.

22. Russian olive (*Elaeagnus angustifolia*) is another ornamental, drought-tolerant, invasive small tree from western Asia that grows in many of the same habitats as tamarisk.

23. Carlson, "The Politics of a Tree," 297.

24. Cameron H. Douglass, Scott J. Nissen, and Charles R. Hart, "Tamarisk Management: Lessons and Techniques," in Sher and Quigley, *Tamarix*, 343.

25. Dan Bean, Tom Dudley, and Kevin Hultine, "Bring on the Beetles!," in Sher and Quigley, *Tamarix*, 380–81.

26. Ibid., 393; and Steve Byrns, "Imported 'Bio-Beetles' Attack Invasive Saltcedar," *AgriLife Today*, August 24, 2012, http://today.agrilife.org/2012/08/24/imported-bio-beetles-attack-invasive-saltcedar/.

27. Mark K. Sogge, Eben H. Paxton, and Charles van Riper III, "Tamarisk in Riparian Woodlands: A Bird's Eye View," in Sher and Quigley, *Tamarix*, 191–92.

28. C. Jack DeLoach, Phil A. Lewis, John C. Herr, Raymond I. Carruthers, James L. Tracy, and Joye Johnson, "Host Specificity of the Leaf Beetle, *Diorhabda elongata deserticola* (Coleoptera: Chrysomelidae) from Asia, a Biological Control Agent for Saltcedars (*Tamarix*: Tamaricaceae) in the Western United States," *Biological Control* 27 (2003): 120.

29. Carlson, "The Politics of a Tree," 300.

30. Sogge et al., "Tamarisk in Riparian Woodlands," 201.

31. Ibid.

32. DeLoach et al., "Host Specificity," 120.

33. Sogge et al., "Tamarisk in Riparian Woodlands," 198.

34. Ibid.

Chapter 6

1. See Margaret R. Slater, "Understanding Issues and Solutions for Unowned, Free-Roaming Cat Populations," *Journal of the American Veterinary Medical Association* 225, no. 9 (2004): 1350–54.

2. Alley Cat Allies, accessed May 10, 2013, http://www.alleycat.org/latest-news/?gclid=CMaJrPStitACFQUIaQodCWwEJA.

3. See Natalie Schutler, "Taking No-Kill Approach with Feral Cat Population," *New York Times*, April 15, 2014, http://www.nytimes.com/2014/04/16/nyregion/a-no-kill-approach-to-feral-cat-control.html?_r=0.

4. Brian Monk, "Open Mic: A Veterinarian's Perspective on the Feral Cat Issue,"

American Birding Association (blog), March 27, 2013, http://blog.aba.org/2013/03/open-mic-addressing-the-feral-cat-issue.html.

5. Feral Cat Awareness Team, "How Does TNR Work?," accessed September 12, 2017, http://www.feralcatawareness.com/tnr.html.

6. Claudio Ottoni, Wim Van Neer, Bea De Cupere, Julien Daligault, Silvia Guimaraes, Joris Peters, Nikolai Spassov, et al., "The Palaeogenetics of Cat Dispersal in the Ancient World," *Nature, Ecology & Evolution* 1 (2017): 1–2.

7. For a sample from North America, see Kerrie Anne T. Loyd, Sonia M. Hernandez, John P. Carroll, Kyler J. Abernathy, and Greg J. Marshall, "Quantifying Free-Roaming Domestic Cat Predation Using Animal-Borne Video Cameras," *Biological Conservation* 160 (2013): 183–89.

8. IUCN, Global Invasive Species Database, Species profile: *Felis catus*, 2016, http://www.issg.org/database/species/ecology.asp?si=24&fr=1&sts=&lang=ENN.

9. Ibid.; and Felix M. Medina, Elsa Bonnaud, Eric Vidal, Bernie R. Tershy, Erika S. Zavaleta, C. Josh Donlan, Bradford S. Keitt, et al., "A Global Review of the Impacts of Feral Cats on Island Endangered Vertebrates," *Global Change Biology* 17, no. 11 (2011): 3503–10.

10. Alley Cat Allies, accessed June 2013, http://www.alleycat.org/page.aspx?pid=325.

11. Texas Parks and Wildlife, Conservation Outreach Program, Issue Briefing Paper, "Management of Feral Cat Colonies," June 2014, 2, http://abcbirds.org/wp-content/uploads/2015/10/Texas-Parks-and-Wildlife-Department-Feral-Cat-Briefing-Paper.pdf.

12. *Neighborhood Cats TNR Handbook: The Guide to Trap-Neuter-Return for the Feral Cat Caretaker*, 2nd ed. ([New York]: Neighborhood Cats, 2013).

13. Maryann Mott, "US Faces Growing Feral Cat Problem," *National Geographic News* September 7, 2004, http://news.nationalgeographic.com/news/2004/09/0907_040907_feralcats.html.

14. Patrick Foley, Janet E. Foley, Julie K. Levy, and Terry Paik, "Analysis of the Impact of Trap-Neuter-Return Programs on Populations of Feral Cats," *Journal of the American Veterinary Association* 227, no. 11 (2005): 1775–81, https://www.avma.org/News/Journals/Collections/Documents/javma_227_11_1775.pdf.

15. Scott R. Loss, Tom Will, and Peter P. Marra, "The Impact of Free-Ranging Domestic Cats on Wildlife of the United States," *Nature Communications* 4, no. 1396 (January, 2013): 1.

16. American Bird Conservancy, "Cats Indoors," accessed July 12, 2014, https://abc-birds.org/program/cats-indoors/cats-and-birds/.

17. "What Is PETA's Stance on Programs That Advocate Trapping, Spaying and Neutering, and Releasing Feral Cats," accessed April 10, 2018, http://www.peta.org/about-peta/faq/what-is-petas-stance-on-programs-that-advocate-trapping-spaying-and-neutering-and-releasing-feral-cats/#ixzz31ubaLMIB; and "Feral Cats: Trapping Is the Kindest Solution," accessed April 10, 2018, http://www.peta.org/issues/companion-animal-issues/companion-animals-factsheets/feral-cats-trapping-kindest-solution/#ixzz31uYq6uea.

18. Centers for Disease Control and Prevention, "Parasites—Toxoplasmosis," August 25, 2017, https://www.cdc.gov/parasites/toxoplasmosis/gen_info/pregnant.html.

19. Ker Than, "Cat Parasite Affects Human Culture," *Livescience*, August 3, 2006, http://www.livescience.com/933-study-cat-parasite-affects-human-culture.html; J. P.

Dubey and J. L Jones, "*Toxoplasma gondii* Infection in Humans and Animals and in the US," *International Journal of Parasitology* 38, no. 11 (2008): 1257–78; Astrid M. Tentner, A. R. Heckeroth, and L. M. Weiss, "*Toxoplasma gondii*: From Animals to Humans," *International Journal of Parasitology* 30, no. 12–13 (2000): 1217–58; and Karen Weintraub, "A Cat Concern," *New York Times*, June 11, 2017.

20. Cornell University, College of Veterinary Medicine, Cornell Feline Health Center, accessed May 17, 2014, http://www.vet.cornell.edu/fhc/Health_Information/topics.cfm.

21. M. Nils Peterson, Brett Hartis, Shari Rodriguez, Matthew Green, and Christopher A. Lepczyk, "Opinions from the Front Lines of Cat Colony Management Conflict," *PLOS One* 7, no. 9 (2012), e44616, doi:10.1371/journal.pone.0044616.

22. Paul L. Barrows, "Professional, Ethical, and Legal Dilemmas of Trap-Neuter-Release," *Journal of the American Veterinary Medical Association* 225, no. 9 (2004): 1365–69.

23. See Beverly L. Garland, "Prowling the Nature-Culture Borderlands: A Geography of the Feral Cat" (master's thesis, University of Texas at Austin, 2007).

Chapter 7

1. Texas A&M, AgriLife Extension, "History of the Red Imported Fire Ant," Texas Imported Fire Ant Research and Management Project, accessed October 31, 2016, http://fireant.tamu.edu/learn/history-of-the-red-imported-fire-ant/; and "Other Impacts of Fire Ants," eXtension, accessed March 27, 2017, http://articles.extension.org/pages/11005/other-impacts-of-fire-ants.

2. Christine Hauser, "Tiny Refugees Hang Together," *New York Times*, August 17, 2017.

3. Rob M. Plowes, research scientist, Department of Integrative Biology, College of Natural Sciences, Brackenridge Field Laboratory, University of Texas at Austin, pers. comm. March 13, 2014; William Vitka, "Nursing Home Ant-Bite Death," *CBS News*, March 12, 2005, http://www.cbsnews.com/news/nursing-home-ant-bite-death-payout/; and Katie Moisse, "Georgia Woman Dies from Fire Ant Sting," *ABC News*, July 19, 2013, http://abcnews.go.com/Health/georgia-woman-dies-fire-ant-sting/story?id=19706086.

4. Anne-Marie Callcott and Homer L. Collins, "Invasion and Range Expansion of Imported Fire Ants," *Florida Entomologist* 79, no. 2 (1996): 240–51.

5. Joshua B. Buhs, *The Fire Ant Wars* (Chicago: University of Chicago Press, 2004), 47–48, 50.

6. Edward G. LeBrun, John Abbott, and Lawrence E. Gilbert, "Imported Red Crazy Ant Displaces Imported Fire Ant," *Biological Invasions* 15 (2013): 2429–42.

7. Matthew C. Fitzpatrick, Jake F. Weltzin, Nathan J. Sanders, and Robert R. Dunn, "The Biogeography of Prediction Error: Why Does the Introduced Range of the Fire Ant Over-Predict Its Native Range?," *Global Ecology and Biogeography* 16, no. 1 (2007): 31, 24–33; and Sanford D. Porter and Dolores A. Savignano, "Invasion of Polygene Fire Ants Decimates Native Ants and Disrupts Arthropod Community," *Ecology* 71, no. 6 (1990): 2095–2106.

8. University of California Davis, Integrated Pest Management Program Online, "Red Imported Fire Ant," 2013, http://www.ipm.ucdavis.edu/PMG/PESTNOTES/pn7487.html; and Louisiana State University, AgCenter Research and Extension, *Managing Imported Fire Ants in Urban Areas*, Publication No. 2187 (2000): 18.

9. Australian Government, Department of the Environment, "Red Imported Fire

Ant," accessed February 12, 2014, http://www.environment.gov.au/biodiversity/invasive-species/insects-and-other-invertebrates/tramp-ants/red-imported-fire; Australian Government, Department of the Environment, "The Reduction of the Biodiversity of Australian Native Fauna and Flora due to the Red Imported Fire Ant," accessed February 12, 2014, http://www.environment.gov.au/node/1458; and Queensland Government, Department Agriculture, Fisheries and Forestry, "Fire Ants," November 2014, https://www.daff.qld.gov.au/plants/weeds-pest-animals-ants/invasive-ants/fire-ant.

10. Kate Creedon, "Race against Time to Quarantine Sydney Outbreak of Red Fire Ants," *Ninenews*, Sydney, Australia, December 4, 2014, http://www.9news.com.au/national/2014/12/04/17/25/deadly-fire-ants-found-in-sydney; and Josh Dye, "Red Imported Fire Ant Outbreak," *Sydney Morning Herald*, December 19, 2014, http://www.smh.com.au/environment/red-imported-fire-ant-outbreak-600-homes-to-be-searched-20141217-129mar.html.

11. University of Texas at Austin, "Fire Ant Project," accessed February 12, 2014, http://www.sbs.utexas.edu/fireant/index.html; and Rob M. Plowes, pers. comm., 2014. There is an ongoing debate regarding whether and to what extent RIFAs attack other ant populations, native or not, and reduce them. There is no doubt that disturbance promotes expansion.

12. Bastiaan M. Drees and David Oi, "Natural Enemies of Fire Ants" (with interactive map of phorid flies), Texas AgriLife Extension, April 4, 2015, http://www.extension.org/pages/30546/natural-enemies-of-fire-ants#.VRWWRmbDs7A.

13. USDA, Agricultural Research Service, "Controlling Red Imported Fire Ants Two Ways," *Science Daily* 21 (August 2009), www.sciencedaily.com/releases/2009/07/090719193246.htm.

14. Walter R. Tschinkel, *The Fire Ants* (Cambridge, MA: Belknap, 2006); and Walter R. Tschinkel and J. R. King, "The Role of Habitat in the Persistence of Fire Ant Populations," *PLOS One* 8, no. 10 (2013): e78580, doi:10.1371/journal.pone.0078580.

15. Robert M. Plowes, John G. Dunn, and Lawrence E. Gilbert, "The Urban Fire Ant Paradox: Native Fire Ants Persist in an Urban Refuge While Invasive Fire Ants Dominate Natural Habitats," *Biological Invasions* 9, no. 7 (2007): 825–36; and Edward G. LeBrun, Robert M. Plowes, and Lawrence E. Gilbert, "Imported Fire Ants near the Edge of Their Range," *Journal of Animal Ecology* 81, no. 4 (2012): 884–95,.

16. Edward G. LeBrun, John Abbot, and Lawrence E. Gilbert, "Imported Crazy Ant Displaces Imported Fire Ant," *Biological Invasions* 15, no. 11 (2013): 2429–42.

17. Ibid., 2440n2; and Marty Toohey, "Crazy Ants Are the New Fire Ants (and Possibly Worse)," *Austin-American Statesman*, December 4, 2016.

18. University of Texas at Austin, "Invasive 'Crazy Ants' Are Displacing Fire Ants in Areas throughout Southeastern US," *ScienceDaily*, May 16, 2013, http://www.sciencedaily.com/releases/2013/05/130516123916.htm.

19. Dina F. Maron, "The Rise of the Crazy Ants," *Scientific American*, February 13, 2014, http://www.scientificamerican.com/article/the-rise-of-the-crazy-ants/; and Toohey, "Crazy Ants Are the New Fire Ants."

20. Barry Bolten, "Species: *Wasmannia auropunctata*," AntWeb, accessed March 27, 2015, www.antweb.org/description.do?rank=species&genus=wasmannia&name=auropunctata; Secretariat of the Pacific Regional Environment Programme, *Managing the Impacts of the*

Little Fire Ants in French Polynesia (Apia, Samoa: SPREP, 2014), http://www.littlefireants.com/LittleFireAnts_ENG.compressed.pdf; James K. Wetterer, "Worldwide Spread of the Little Fire Ant," *Terrestrial Arthropod Reviews* 6 (2013): 173–84; CABI, Invasive Species Compendium, "Little Fire Ant," accessed April 14, 2018, https://www.cabi.org/isc/datasheet/56704; and Rob Plowes, pers. comm., April 16, 2018.

21. Kate Kelland, "Cancer Agency Left in the Dark," *Reuters*, June 14, 2017, http://www.reuters.com/investigates/special-report/glyphosate-cancer-data/.

Chapter 8

1. Nina Leigh Vizcarrando, *A Kind of Wild* (master's thesis [film], University of Texas at Austin, 2014), 3 [script].

2. Charly Seale, "Can Hunting Endangered Animals Save the Species?," *CBS News 60 Minutes*, January 30, 2012, http://www.cbsnews.com/news/can-hunting-endangered-animals-save-the-species.

3. T. Baccus, "Impacts of Game Ranching," *Transactions of the North American Wildlife and Natural Resources Conference* 67 (2002): 276–88.

4. Charles W. Ramsey, "State Views of Governmental and Private Programs," in *Introductions of Exotic Animals* (College Station, TX: Caesar Kleberg Research Program, 1967), 9–10; Michael Bowlin, "Patio Ranch Once Boasted 'Blind Service' for Hungry Hunters," *Kerrville Daily Times*, April 12, 1992, 3; and "The Patio Ranch: Tell a Different Hinting Story," accessed April 14, 2018, www.thepatioranch.com/about-the-patio-ranch-texas-hunting.html.

5. George W. Cox, *Alien Species in North America and Hawaii* (Washington, DC: Island Press, 1999), 194–95.

6. Exotic Wildlife Association, accessed July 10, 2016, http://www.myewa.org/index.cfm; and Ramsey, "State Views," 9–10.

7. Gregory L. Butts, "The Status of Exotic Big Game in Texas," *Rangelands* 1, no. 4 (1979): 152–53.

8. Max Traweek and Roy Welch, "Exotics in Texas," Texas Parks and Wildlife Department, April 1992, http://tpwd.texas.gov/publications/pwdpubs/media/pwd_bk_w7000_0206.pdf.

9. Rusty Middleton, "Texotics," *Texas Parks and Wildlife*, April 2007, http://tpwmagazine.com/archive/2007/apr/ed_3/.

10. Stephen Demaris, David A. Osborn, and James J. Jackley, "Exotic Big Game: A Controversial Resource," *Rangelands* 12, no. 2 (1990): 121–25.

11. Middleton, "Texotics."

12. Keith A. Clark, R. M. Robinson, R. G. Marburger, L. P. Jones, and J. H. Orchard, "Malignant Catarrhal Fever in Texas Cervids," *Journal of Wildlife Diseases* 6, no. 4 (1970): 376–83.

13. R. M. Robinson, L. P. Jones, T. J. Galvin, and G. M. Harwell, "Elaeophorosis in Sika Deer in Texas," *Journal of Wildlife Diseases* 14, no. 1 (1978): 137–41; and John M. Tomaček, Terry Hensley, Walt E. Cook, and Bob Dittmar, *A Guide to Chronic Wasting Disease (CWD) in Texas Cervids* (College Station, TX: AgriLife Extension Service, 2015).

14. Lynn Brezosky, "Antelopes Go from Pests to Plague in Texas," *Los Angeles Times*, April 29, 2007; and Steve Byrns, "Texas Cattle Fever Ticks Are Back,"

AgriLife Today, February 2, 2017, https://today.agrilife.org/2017/02/02/texas-cattle-fever-ticks-back-vengeance/.

15. Jan Reid, "Bring 'Em Back Alive," *Texas Monthly* 13, no. 3 (1985): 134–37, 217–21, esp. 219; and Robert Strickland, "Tick Identification," *Proceedings of Seminar on Tick Eradication Measures*, Puerto Rico, September 3–6, 1985, 2, https://books.google.com/books?id=uPgOAQAAIAAJ.

16. For web chat room comments, see Texas Hunting Forum, accessed September 19, 2010, http://texashuntingforum.com/forum/ubbthreads.php/topics/1687826/Re_Looking_for_Oryx_Hybrids_to.

17. IUCN SSC Antelope Specialist Group, *Oryx dammah*, IUCN Red List of Threatened Species, 2016, e.T15568A50191470, http://dx.doi.org/10.2305/IUCN.UK.2016-2.RLTS.T15568A50191470.en.

18. Jim Forsyth, "Out of Africa," *Reuters*, May 19, 2015, http://www.reuters.com/article/us-usa-rhino-texas-idUSKBN0O418P20150519; and Dan Solomon, "Up to 4 Percent of Africa's Rhinos Might Relocate to South Texas," *Texas Monthly*, May 6, 2015, http://www.texasmonthly.com/the-daily-post/up-to-4-percent-of-africas-rhinos-might-relocate-to-south-texas/.

19. IUCN SSC Antelope Specialist Group, *Antilope cervicapra*, IUCN Red List of Threatened Species, 2017, e.T1681A50181949, http://dx.doi.org/10.2305/IUCN.UK.2017-2.RLTS.T1681A50181949.en; and William B. Davis and David J. Schmidly, "Blackbuck," in *The Mammals of Texas—Online Edition* (Lubbock: Texas Tech University, 1997), http://www.nsrl.ttu.edu/tmot1/anticerv.htm. Details of the 10 animals introduced in 1970 can be found in Virginia Kraft, "Bucking Up the Pakistanis," *Vault*, April 27, 1970, https://www.si.com/vault/1970/04/27/610849/bucking-up-the-pakistanis.

20. Elizabeth Cary Mungall, *Exotic Animal Field Guide* (College Station: Texas A&M University Press, 2007).

Chapter 9

1. Herbarium specimens and comments on distribution are based on collections housed at the University of Texas at Austin's Billie L. Turner Plant Resources Center.

2. Richard N. Mack, "The Commercial Seed Trade: An Early Dispenser of Weeds in the US," *Economic Botany* 45, no. 2 (1991): 267; and Julia F. Morton, "Brazilian Pepper: Its Impact on People, Animals and the Environment," *Economic Botany* 32, no. 4 (1978): 353–55.

3. J. P. Cuda, A. P. Ferriter, V. Manrique, and J .C. Medal, eds., *Florida's Brazilian Peppertree Management Plan*: *Recommendations from the Brazilian Peppertree Task Force*, 2nd ed. (Gainesville: Florida Exotic Pest Plant Council, 2006), 6, http://www.fleppc.org/Manage_Plans/BPmanagPlan06.pdf; S. D. Hight, J. P. Cuda, and J. C. Medal, "Brazilian Peppertree," in *Biological Control of Invasive Plants in the Eastern United States*, ed. Roy Van Driesche, Suzanne Lyon, Bernd Blossey, Mark Hoddle, and Richard Reardon, Publication FHTET-2002-04 (Morgantown, WV: US Department of Agriculture, Forest Service, Forest Health Technology Enterprise Team, 2002), 311; Mack, "Commercial Seed Trade," 267; and Amy Ferriter, ed., *Brazilian Pepper Management Plan for Florida* (Gainesville: Florida Exotic Pest Plant Council, 1997), 2, 12.

4. Based on herbarium specimens housed at the University of Texas at Austin's Billie

Lee Turner Plant Resource Center; and minutes of the Texas Gulf Region Cooperative Weed Management Area, TexasInvasives.org, February 24, 2014, http://www.texasinvasives.org/professionals/cwma/tgrcwma/Feb24_Minutes.pdf.

5. Dean A. Williams, William A. Overholt, James P. Cuda, and Colin R. Hughes, "Chloroplast and Microsatellite DNA Diversities Reveal the Introduction History of Brazilian Peppertree (*Schinus terebinthifolius*) in Florida," *Molecular Ecology* 14 (2005): 3652–53.

6. Rachelle Meyer, "*Schinus terebinthifolius*," in *Fire Effects Information System*, US Department of Agriculture, Forest Service, Rocky Mountain Research Station, Fire Sciences Laboratory, 2011, http://www.fs.fed.us/database/feis/; and Cuda et al., *Florida's Brazilian Peppertree*, 26.

7. Morton, "Brazilian Pepper," 355.

8. Spencer C. H. Barrett, "Waterweed Invasions," *Scientific American* 260 (1989): 97.

9. M. H. Julien, T. D. Center, and P. W. Tipping, "Floating Fern (*Salvinia*)," in Van Driesche et al., *Biological Control of Invasive Plants in the Eastern United States*, 21; Barrett, "Waterweed Invasions," 96; and Colette Jacono and Bob Pitman, "*Salvinia molesta*: Around the World in 70 Years," *Aquatic Nuisance Species Digest* 4, no 2 (2001): 14.

10. I. W. Forno and A. S. Bourne, "Studies in South America of Arthropods on the *Salvinia auriculata* Complex of Floating Ferns and Their Effects on *S. molesta*," *Bulletin of Entomological Research* 74 (1984): 619; and Julien et al., "Floating Fern," 18.

11. D. Johnson, "Giant Salvinia Found in South Carolina," *Aquatics* 17 (1995): 22; and Colette C. Jacono, "*Salvinia molesta* (Salviniaceae), New to Texas and Louisiana," *Sida* 18, no. 3 (1999): 927.

12. The Ramsar Sites, accessed December 2, 2015, http://www.ramsar.org/sites-countries/the-ramsar-sites; and Greater Caddo Lake Association of Texas, *Fear No Weevil*, video, version 2, accessed March 24, 2015, http://gclaoftx.com/.

13. Daren Horton, *Fighting Giant Salvinia: High-Production Weevil Greenhouse Facility at Caddo Lake*, video, October 23, 2014, http://caddosalvinia.blogspot.com/2014_10_01_archive.html.

14. Abhishek Mukherjee, Allen Knutson, Daniel A. Hahn, and Kevin M. Heinz, "Biological Control of Giant Salvinia (*Salvinia molesta*) in a Temperate Region: Cold Tolerance and Low Temperature Oviposition of *Cyrtobagous salviniae*," *BioControl* 59 (2014): 788–89; and Lucas Gregory, *Giant Salvinia Weevil Research at Caddo Lake*, parts 1 and 2, video, April 9, 2014, http://caddosalvinia.blogspot.com/.

15. Mark Wilson, "Zebra Mussels Discovered in Lake Austin," *Austin American-Statesman*, August 17, 2017; and Mark Wilson, "Invasive Zebra Mussels Infesting Lake Travis," *Austin American-Statesman*, June 27, 2017.

16. Michael Young, "Water from Lake Texoma," *Dallas News*, May 28, 2014, http://www.dallasnews.com/news/texas/2014/05/28/water-from-lake-texoma-could-be-back-in-north-texas-homes-soon.

17. Marrone Bio Innovations, "Zebra Mussel-Killing Bacteria Could Help Native Species in the Great Lakes," accessed April 5, 2017, http://marronebioinnovations.com/zebra-mussel-killing-bacteria-could-help-native-species-in-the-great-lakes/; and Robert Boyle, "Science Tales on a Silent Invader," *New York Times*, February 24, 2014.

18. Melissa Gaskill, "Roar of the Lionfish," *Texas Parks and Wildlife*, December 2013,

https://tpwmagazine.com/archive/2013/dec/ed_3_lionfish/; and Melissa Gaskill, "As Lionfish Invade the Caribbean and Gulf of Mexico, Conservationists Say Eat Up," *Scientific American*, December 11, 2013, https://www.scientificamerican.com/article/does-eating-lionfish-work/.

19. NOAA, Office of Marine Sanctuaries, "Flower Garden Banks Marine Sanctuary," 2017, https://nmsflowergarden.blob.core.windows.net/flowergarden-prod/media/archive/document_library/aboutdocs/fgbnmsflyer.pdf; and NOAA, Office of Marine Sanctuaries, "Flower Garden Banks, Invasive Lionfish," 2015, http://flowergarden.noaa.gov/education/invasivelionfish.html.

20. Quoted in Gaskill, "Roar of the Lionfish."

21. Lad Akins, *Lionfish Cookbook* (Key Largo, FL: REEF, 2012).

22. Leslie D. Hartman notes that an angler took a lionfish from Packery Channel jetties near Corpus Christi in July 2013. He reported other specimens in coastal waters. See "Lionfish Stalking Our Bays?," *Texas Saltwater Fishing Magazine*, November 2013.

23. Christopher Clarey, "As the America's Cup Team Prepare," *New York Times*, April 18, 2017.

24. Andrew B. Barbour, Michael S. Allen, Thomas K. Frazer, Krista D. Sherman, "Evaluating the Potential Efficacy of Invasive Lionfish (*Pterois volitans*) Removals," *PLOS One* 6, no. 5 (2011), http://journals.plos.org/plosone/article?id=10.1371/journal.pone.0019666; Lee Sausley, "Texas vs. Lionfish," February 7, 2016, http://www.kristv.com/story/31161698/texas-vs-lionfish; and Raven Walker, "Marine Biologist," *Texas A&M Today*, 2016, http://today.tamu.edu/2016/08/09/marine-biologist-eat-lionfish-before-they-eat-up-the-gulf-coast/.

25. Gary L. Parsons, "Emerald Ash Borer: A Guide to Identification and Comparison to Similar Species," November 2008, http://www.emeraldashborer.info/documents/eab_id_guide.pdf.

26. The Emerald Ash Borer Information Network (www.emeraldashborer.info) is a multinational, multiuniversity site loaded with information, maps, and an extensive webinar series that cover the insect's basic biology, history, and management strategies.

27. G. M. Lovett, M. Weiss, A. M. Liebhold, T. P. Holmes, B. Leung, K. F. Lambert, D. A. Orwig, et al., "Nonnative Forest Insects and Pathogens in the United States: Impacts and Policy Options," *Ecological Applications* 26, no. 5 (2016): 1443.

28. Texas A&M Forest Service, "Newsroom: Texas A&M Forest Service Confirms Emerald Ash Borer Found in Texas," May 23, 2016, http://texasforestservice.tamu.edu/content/article.aspx?id=24246; and Shane Harrington, Texas A&M Forest Service, pers. comm., December 5, 2017.

29. Don Cipollini, "White Fringetree as a Novel Larval Host for Emerald Ash Borer," *Journal of Economic Entomology* 108, no. 1 (2015): 370.

30. Leah S. Bauer, Jian J. Duan, Juli R. Gould, and Roy Van Driesche, "Progress in the Classical Biological Control of *Agrilus planipennis* Fairmaire (Coleoptera: Buprestidae) in North America," *Canadian Entomologist* 147 (2015): 300, 311–12; and US Department of Agriculture, Forest Service, "Development of Novel Ash Hybrids," April, 3, 2015, https://www.nrs.fs.fed.us/disturbance/invasive_species/eab/control_management/novel_ash_hybrids/.

31. See the Nature Conservancy, "Firewood—Buy It Where You Burn It," accessed

April 14, 2018, https://www.nature.org/ourinitiatives/urgentissues/land-conservation/forests/firewood-buy-it-where-you-burn-it.xml; and "Protect the Trees You Love from Tree-Killing Bugs," accessed April 14, 2018, https://www.dontmovefirewood.org/.

32. See "A Co-ordinated Response to the Devastating Bat Disease," White-nose Syndrome.org, accessed April 5, 2018, https://www.whitenosesyndrome.org/; Texas Parks and Wildlife, "White-nose Syndrome," accessed April 5, 2018, https://tpwd.texas.gov/huntwild/wild/diseases/whitenose/; Texas Parks and Wildlife, *White-nose Syndrome Action Plan*, February 2017, http://tpwd.texas.gov/huntwild/wild/diseases/whitenose/docs/TPWD_WNS_Plan.pdf; and Katie Hall, "Fungus That Can Kill Bats Hits Area," *Austin-American Statesman*, April 5, 2018, B1, 3.

33. Texas Parks and Wildlife, *Action Plan*, 4–5. As 12 of the bat species in Texas are regular hibernators, it does not look promising that the syndrome will not break out.

34. Ibid.

Conclusion

1. Bruce A. Stein and Stephanie R. Flack, eds. *America's Least Wanted* (Arlington, VA: Nature Conservancy, 1996), 7.

2. Brij Gopal and K. P. Sharma, *Water-Hyacinth* (Eichhornia crassipes)*, the Most Troublesome Weed of the World* (Delhi: Hindasia, 1981).

3. iNaturalist.org, "Spotted Knapweed," June 7, 2014, http://www.inaturalist.org/observations/729612.

4. US Department of Agriculture, National Institute of Food Agriculture, "Biological Control Program," accessed April 13, 2018, https://nifa.usda.gov/program/biological-control-program.

5. John R. Meyer, "Pest Control Tactics," November 4, 2003, https://projects.ncsu.edu/cals/course/ent425/text19/biocontrol.html.

6. Matthew J. W. Cock, Joop C. van Lenteren, Jacques Brodeur, Barbara I. P. Barratt, Franz Bigler, Karel Bolckmans, Fernando L. Cônsoli, et al., "Do New Access and Benefit Sharing Procedures under the Convention on Biological Diversity Threaten the Future of Biological Control?," *BioControl* 55, no. 2 (2010): 199–218.

7. Carl Zimmer, "Gene-Editing Strategy," *New York Times*, November 20, 2017.

8. Christina Procopiou, "Get Those Pests," *Wildflower Magazine*, March 3, 2014, https://www.wildflower.org/magazine/conservation/get-those-pests.

9. City of Austin, Texas, *Invasive Species Management Plan*, 2013, 7–8, http://austintexas.gov/sites/default/files/files/Watershed/invasive/COA-ISMP-Final-7-11-12.pdf.

10. John Davis, wildlife diversity leader, Texas Parks and Wildlife Department, pers. comm., August 1, 2014.

11. Ibid.

12. Jim Rogers, "My Sand County Journey: A Private Texas Landowner's Insights, Experiences, Advice and Adventures," *L.I.P. Bulletin* 8 (2016): 5–7, tpwd.texas.gov/publications/pwdpubs/media/pwd_lf_w7000_1405_2016.pdf.

13. US Fish and Wildlife Service, "US Fish and Wildlife Service Evaluating Safe Harbor Agreement for the Houston Toad," press release, August 26, 2006, https://www.fws.gov/news/ShowNews.cfm?ref=u.s.-fish-and-wildlife-service-evaluating-safe-harbor-agreement-for-the-&_ID=35778.

14. Center for Biological Diversity, "Agreement: Monarch Butterflies to Get Endangered Species Protection Decision by 2019," press release, July 5, 2016, https://www.biologicaldiversity.org/news/press_releases/2016/monarch-butter fly-07-05-2016.html.

15. Darryl N. Jones and S. James Reynolds, "Feeding Birds in Our Towns and Cities: A Global Research Opportunity," *Journal of Avian Biology* 39, no. 3 (2008): 265–71; and Mireya Navarro, "Record Number of Peregrine Falcons in New York State," *New York Times*, February 12, 2009, http://www.nytimes.com/2009/02/13/nyregion/13falcon.html.

16. For overviews, see Jaymi Heimbuch, "10 Fascinating Facts about Urban Coyotes," Urban Coyote Initiative, accessed July 5, 2016, http://urbancoyoteinitiative.com/10-fas-cinating-facts-about-urban-coyotes/; and Marissa Fessenden, "Coywolves Are Taking Over," *Smithsonian.com*, November 3, 2015, https://www.smithsonianmag.com/smart-news/coywolves-are-taking-over-eastern-north-america-180957141/.

17. Frances Bonier, Paul R. Martin, and John C. Wingfield, "Urban Birds Have Broader Environmental Tolerance," *Biology Letters* 3, no. 6 (2007): 270–73.

18. Lawrence E. Conole, "Degree of Adaptive Response in Urban Tolerant Birds," *PeerJ* 2 (2014): e306, https://peerj.com/articles/306/.

19. Sherri D. Graves and Arthur M. Shapiro, "Exotics as Host Plants of the California Butterfly Fauna," *Biological Conservation* 110 (2003): 413–33.

20. Helen Briggs, "Urban Habitats 'Provide Haven' for Bees," *BBC News*, Science & Environment, February 11, 2015, http://www.bbc.com/news/science-environment-31359984; Tristan Donovan, *Feral Cities: Adventures with Animals in the Urban Jungle* (Chicago: Chicago Review Press, 2015), 7; "Fox Lived in the Shard Skyscraper at London Bridge," *BBC News*, February 24, 2011, http://www.bbc.com/news/uk-england-london-12573364; and Anne-Sophie Brändlin, "Why Urban Wildlife Is Thriving in Berlin," *DW*, July 7, 2016, http://www.dw.com/en/why-urban-wildlife-is-thriving-in-berlin/a-19386181.

21. Donovan, *Feral Cities*, 6.

22. Adam Cruise, "Is This the End for South Africa's Famed Urban Baboons?," *National Geographic*, April 22, 2016, https://news.nationalgeographic.com/2016/04/160422-baboons-cape-town-conservation-south-africa/; and Richard Conniff, "Learning to Live with Leopards," *National Geographic*, November 10, 2015, http://ngm.nationalgeographic.com/2015/12/leopards-moving-to-cities-text.

23. Kevin M. Anderson, coordinator, Austin Water, Center for Environmental Research, City of Austin, Texas, pers. comm., 2014.

24. City of Austin, *Invasive Species Management Plan*, 37–39.

25. J. Lane, "Overwintering Monarch Butterflies in California, Past and Present," in *Biology and Conservation of the Monarch Butterfly*, ed. Stephen B. Malcom and Myron Zalucki (Los Angeles: Natural History Museum, 1993), 337–38; "Monarch Butterflies in California Need Eucalyptus Trees for Their Winter Roost," *Death of a Million Trees* (blog), November 1, 2013, https://milliontrees.me/2013/11/01/monarch-butterflies-in-california-need-eucalyptus-trees-for-their-winter-roost/.

26. USGS, Patuxent Wildlife Research Center, North American Breeding Bird Survey, 2015, "House Sparrow" and "European Starling," https://www.mbr-pwrc.usgs.gov/cgi-bin/

atlasa15.pl?06882&1&15&csrfmiddlewaretoken=3YKakk7LxT2ki6NSp14mstudYCqdW02C; and Donovan, *Feral Cities*, 49.

27. Molly Morrow, "Austin's Wild Monk Parakeets," *The Austinot*, February 26, 2016, http://austinot.com/monk-parakeets-austin; Peter English, "Austin's Parakeets," *Highland Park West Balcones Area Neighborhood Association* 11, no. 1 (2015): 4, http://5f8c-274712c4ea693cc1-fdbcf82d3dfc08785157cfod6fc8ed50.r16.cf1.rackcdn.com/1501HP.pdf; and CBS Austin, "Parakeet Nest Fire," June 6, 2017, http://cbsaustin.com/news/local/video-parakeet-nest-fire-knocks-out-power-to-1000-austin-energy-customers.

28. For a website dedicated to research on urban parrot populations, see City Parrots, accessed November 4, 2016, http://cityparrots.org/mission/.

29. Noel F. Snyder and Keith Russell, "Carolina Parakeet," in *Birds of North America*, ed. P. G. Rodewald (Ithaca, NY: Cornell Laboratory of Ornithology, 2002), https://birdsna.org/Species-Account/bna/species/carpar/introduction; John J. Audubon, "Carolina Parrot," in *Birds of America*, plate 26, accessed September 28, 2016, http://www.audubon.org/birds-of-america/carolina-parrot; and Daniel McKinley, "History of the Carolina Parakeet in Its Southwestern Range," *Wilson Bulletin* 1.76, no. 1 (1964): 68–93.

Appendix 1

1. USDA, National Agricultural Library, Laws and Regulations, "International Laws and Regulations," last modified May 18, 2016, http://www.invasivespeciesinfo.gov/laws/intl.shtml, provides an introduction to global, regional, and national legal instruments. Each can be accessed and explored.

2. Article 196 of the UN Convention of the Law of the Sea requires states to take measures to control the introduction of species, notably alien ones, to a new marine environment.

3. Convention on Biological Diversity, Conference of Parties Decisions, accessed June 12, 2013, http://www.cbd.int/invasive/cop-decisions.shtml; and Melodie A. McGeoch, Stuart H. M. Butchart, Dian Spear, Elrike Marais, Elizabeth J. Kleynhans, Andy Symes, Janice Chanson, et al., "Global Indicators of Biological Invasion," *Diversity and Distributions: Journal of Conservation Biogeography* 16, no. 1 (2010): 95–108.

4. See Invasive Species Specialist Group, "View 100 of the World's Worst Invasive Alien Species," accessed April 28, 2018, http://www.issg.org/worst100_species.html.

5. See NIIS, accessed April 28, 2018, http://www.niiss.org/cwis438/websites/niiss/About.php?WebSiteID=1.

6. In September 2017, the International Maritime Organizations' International Convention for the Control and Management of Ships' Ballast went into effect, requiring among other actions that all ships implement a ballast-water management plan.

7. Kristina Alexander, "Injurious Species Listings under the Lacey Act: A Legal Briefing," Congressional Research Service, 7-5700, August 1, 2013, http://nationalaglaw-center.org/wp-content/uploads/assets/crs/R43170.pdf.

8. US Fish and Wildlife Service, International Affairs, "The Lacey Act," accessed October 9, 2015, https://www.fws.gov/international/laws-treaties-agreements/us-conservation-laws/lacey-act.html; and USDA, APHIS, "Lacey Act," 2015, https://www.aphis.usda.gov/wps/portal/aphis/ourfocus/planthealth?1dmy&urile=wcm%3apath%3a%-

2Faphis_content_library%2Fsa_our_focus%2Fsa_plant_health%2Fsa_import%2Fsa_lacey_act%2Fct_lacey_act.

9. National Invasive Species Council, *Management Plan 2016–2018* (Washington, DC: National Invasive Species Council, 2016), https://www.doi.gov/sites/doi.gov/files/uploads/2016–2018-nisc-management-plan.pdf; and USDA, National Agricultural Library, "Laws and Regulations," 2016, https://www.invasivespeciesinfo.gov/whats-new/laws-and-regulations.

10. Texas Department of Agriculture, Regulatory Programs, Pest Survey, accessed October 6, 2016, https://texasagriculture.gov/RegulatoryPrograms/PlantQuality/PestSurvey.aspx.

11. For the regulations to administer this statute, see "Noxious and Invasive Plants," Subchapter T, Texas Administrative Code, chap. 19, accessed April 28, 2018, txrules.elaws.us/rule/title4_chapter19_subchaptert.

12. Texas Invasive Species Coordinating Committee Minutes, February 11, 2015, http://www.tiscc.texas.gov/wp-content/uploads/2016/03/February-11-2015-Minutes.pdf; and USDA, National Agricultural Library, Resource Library, "Invasive Species Lists: US, Individual States: Texas," 2017, https://www.invasivespeciesinfo.gov/resources/lists4states.shtml.

13. Texasinvasives.org, accessed October 12, 2015, http://www.texasinvasives.org/action/spreadword.php.

Appendix 2

1. Brenda Kellar, "Honeybees across America," accessed April 28, 2018, http://www.orsba.org/download/Honey%20Bees%20Across%20America.html; and Jefferson Foundation, "Bees and Honey," in *Notes on the State of Virginia,* accessed April 28, 2018, https://www.monticello.org/site/plantation-and-slavery/bees-and-honey.

Index

NOTE: Page numbers in *italic* type denote photographs or tables; those in **bold** denote maps.

Act on, 97–98; introduction to Massachusetts, 95, 208; introduction to Texas, 95; laws naming, 203; on Texas invasive plant list, 204; origin, 94; root growth, 98; shifting opinion regarding, 98–99, 102, 181; soil salinity of, 101; southwestern willow flycatcher and, *106*, 107–108, 152, 195; spread, 1, 93, 97–102, 177, 208; utility, 95–97, 107–108, 180, 181; water use, 101–102

salvinia, giant, 20, 160, *161*, 162–63, 178, 179, 182, 185, 191, 215

Salvinia molesta, 160, **161**

scab, sweet orange, 203

Schieffelin, Eugene, 7, 33, 36, 42

Schinus terebinthifolius, 157, **158**

Schreiner, Charles, III, 143–44

scimbok, 151, 178

sheep: aoudad, 146, 148, *149*; Barbados, 151; Barbary (*see* sheep, aoudad); Corsican, 151, 178; desert bighorn, 149; diversity of, in Texas, 144; Hawaiian black, 151; Iranian red, 151; mouflon, 151; mouflon-Barbados, 146; painted desert, 151; raised with exotic game, 147; Rambouillet, 151; Texas dall, 151; transcaspian urial, 151

silk tree, *182*

snail: channeled apple, 203; giant ramshorn, 213. *See also* melania, red-rimmed

snake, brown tree, 201, 203

Solenopsis invicta, **124**, 125

Solenopsis rickteri, 127

sorrel, 207

sparrow, house: 25; as cuisine, 31; attitudes toward, 7, 27–31, 181; control (trapping), 31, 184; damage, 2, 9, 25, 28, 29–30, 37, 179; endangered in Britain, 195; introduction to New York, 7, 26, 208; introduction to Texas, 26, 31; law against injuring, 27; prohibition under Lacey Act, 27, 203; recent decline of,

32, 195; spread, 26–27, 31–32, 191, 192; threat to purple martin, 23–25, 29–30, 179; utility of, 27, 181

sparrow, English. *See* sparrow, house

"sparrow war," 28–31

squirrel, ground, 210

St. John's wort, 207

star-vine, Indian, 56, 212

starling, common. *See* starling, European.

starling, European: *32*; attitudes toward, 33, 181; cattle feedlots and, 33, 35, 178; control, 35, 184; damage, 2, 9, 25, 33–36; flocking behavior of, *35*; introduction to New York, ix, 7, 33, 210; introduction to Texas, ix, 36, 210; laws naming, 203; prohibition under Lacey Act, 27, 203; spread, 34–35, 177, 191; recent increases and declines of, 195; threat to aircraft, 34; threat to purple martin, 23–25; 36–38, 184

stillingia oil, 77, 89

Sturnus vulgaris, 33

sugarcane, 46, 49, 126, 208

swallow, barn, viii

swine, feral. *See* hog, feral

Tadarida brasiliensis, 175

tallow, Chinese: *75, 77*; absence of predators, 83; abundance, 76, 81–83; as petroleum substitute, 88–90; beekeepers' advocacy for, 88, 180; Benjamin Franklin's introduction of, 78, 207; bird use of, 81, 84–85, 180; changing attitude toward, 81, 91; climate models on, 85–86; control, 86–87; cultivation of, 77, 79, 83, 89–90; damage, 76, 81, 86, 179; distribution, **76;** feral hog spread of, 85; genetic shifts in, 83–84; growth and seed production, *77,* 82–84; honey, 88; introduction to Georgia, 78, 207; introduction to Texas, 79; origin, 78; on Texas invasive

Miller Dam, 59, 164; Town Lake, 196

Texas (natural areas, parks, refuges):
Aransas National Wildlife Refuge,
143; Big Bend National Park, 92; Big
Thicket, 81, 85; Huntsville State Park,
42; Lafitte's Cove Nature Preserve,
48; Lost Maples State Park, xi; Lower
Rio Grande Valley National Wildlife
Refuge, 150; Old Tunnel State Park,
175; Palo Duro Canyon, 149; Zilker
Park, 196

Texas (state): control of Mediterranean
fruit fly, 10; list of noxious and invasive plants, 204; percentage of invasive species in, 18

Texas A&M Extension Service, 188

Texas A&M Forest Service, 160, 172

Texas Agricultural Experiment
Station, 59

Texas Agrilife Extension Service, 64,
150, 204

Texas Department of Agriculture, 16,
90, 204

Texas Gulf Coast Cooperative Weed
Management Area, 160

Texas Highway Department, 79, 207

Texas Invasive Plant and Pest Council
(TIPPC): 205, 218; pre-establishment
conference, 217

Texas Invasive Species Coordinating
Committee (TISCC), 204

Texas Invasive Species Institute, 184

TexasInvasives.org, 183, 184,
186, 204–205

Texas Parks and Wildlife Department:
action plan for white-nose syndrome,
175; as aquatic species monitor, 20;
census of exotic game, 145–46; Conservation Outreach Program, 115; hog
control, 74; invasive species list, 204;
management plan for red lionfish,
169; permit for carp, 57, 214; policy
on zebra mussels, 167; Wildlife and
Inland Fisheries Division, 188; Wild-

life Diversity Program, 187

Texotics. See game, exotic

tilapia: 4; blue, 213

tinamou, 145

toad: cane, 212; horny (see lizard,
horned); Houston, 189

Toxoplasma gondii, 118

tree of heaven, 182

Triadica sebifera, 76, 77, 82, 83

Trichechus manatus, 49

trout, rainbow, 12, 215

turkey, 6, 144, 188, 206

United Nations: Convention on the Law
of the Sea, 201; Food and Agriculture
Organization, 201, 217

United States Army Corps of Engineers,
40, 95, 165, 203, 208

United States Bureau of Biological
Survey, 36

United States Department of Agriculture: Agricultural Research Service,
103–104; Animal and Plant Health
Inspection Service (APHIS), 74, 105,
107, 217; battle with ticks, 150; Bureau
of Plant Industry, 7, 41, 79, 83, 145,
207, 210, 211; Division of Wildlife
Services, 65; Environmental Quality
Incentives Program (EQIP), 189; Foreign Plant Introduction Division, 181;
growing saltcedar, 95; invasive species profiles, 218; Natural Resources
Conservation Service, 188, 189; origin
in Patent Office, 208; Section of Seed
and Plant Introduction, 7, 96, 210;
survey of red imported fire ant, 128

United States Environmental Protection
Agency (EPA), 71, 202, 203

United States Fish and Wildlife Service
(USFWS): assistance with invasive
species, 188; concerns with saltcedar
beetles, 107; Foreign Game Program,
145; listing of carp, 218; listing of
invasive species, 203; Nonindigenous